THE BOOK OF COUNTRY CRAFTS

The Book of Country Crafts

by

RANDOLPH WARDELL JOHNSTON

On Working with Wood, Clay, Metals, Stone, and Color, with Many New Recipes and Secrets of the Crafts

South Brunswick and New York:
A. S. Barnes and Company
London: Thomas Yoseloff Ltd

Library of Congress Catalogue Card Number: 64-19168

A. S. Barnes and Co., Inc.
Cranbury, New Jersey 08512

Thomas Yoseloff Ltd
108 New Bond Street
London W1Y OQX, England

SBN: 498 06127 2
Printed in the United States of America

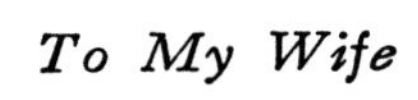

To My Wife

Preface

This is not a text-book, nor is it, as many text-books are, a scholarly compilation of material from many other books. I am not a bookish man, but on the contrary have always had more fun doing things than reading of, or writing about them. However, when I have had to consult books for my work, I have not infrequently found them compilations made from many unverified sources. Because some writers of books are not conversant with the practice of the craft in question, they will not notice omissions of important details in the original sources, and often they will unknowingly transcribe these errors into their work. It is my aim to avoid such shortcomings in my book by writing only of things of which I have personal knowledge. I have always had a great curiosity about crafts and processes of all sorts. Wherever I go I ask many questions and endeavour always to understand the principles. My environment and circumstances have been particularly fortunate for the acquisition of considerable useful knowledge.

As the son of an inventor and engineer, I grew up among tools and machines of all kinds. My school training was designed to prepare me for the profession of architecture, which as a draughtsman and helper at various trades, I followed for awhile. Then I found that sculpture was more interesting and pro-

Preface

ceeded to study that at a number of schools, including the Central School of Arts and Crafts in London. A great lover of thoroughness, I next went to a Medical School to learn anatomy by dissection. In the years that have passed since then, I have not only worked as a sculptor in wood, stone, bronze and terra-cotta, but have also designed and supervised the making of pottery, machinery, small buildings, kilns, melting furnaces, stage-settings and properties.

There are therefore a number of widely differing crafts in each of which, while I am not a master, I can produce results. It is my aim to put tools and materials in your hands that you too may produce results. For the development of genuine mastery in any one field you will naturally wish to read many books and see the work of many masters. Nevertheless, the worker who, living in country districts, is cut off from ready access to libraries, museums and to the work shops of other craftsmen, will find in this book sufficient of solid basic information to make a good beginning and to go a long way towards developing his own style, without any other help. Moreover, in some parts of the book he will find useful formulae and processes which are the result of years of research and experiment on my part and have never been published before.

Limitation of space suggests the emphasis on material not dealt with in other books. Weaving, for example, is omitted because it is completely covered by Mary Meigs Atwater's THE SHUTTLE-CRAFT

Preface

BOOK OF AMERICAN HAND-WEAVING. *Other interesting craftbooks are noted in the bibliography.*

Inseparable from good craftsmanship is the development of style and quality in design. The first requirements are a ready acceptance of the peculiar qualities and limitations of the material to be handled and a fearless facing and mastering of the difficulties of handling it. One might say that design is nothing more than taste and a sense of fitness applied through a feeling for materials to a particular problem. As for this sympathetic feeling for materials, it is of the utmost importance. The true craftsman understands that materials, like people, have the defects of their qualities. Thus knowing that glass, for instance, has the quality of extreme hardness he does not demand from it also the properties of flexibility and ductility as he does from copper, but accepts its brittleness as being complementary to its character of hardness. So with all materials—he does not attempt to make a bronze statue look like a clay one, nor a clay dish like a wicker basket, but always exploits and expresses the qualities inherent in the materials themselves.

Joy in working begets good work and doing good work in a handicraft is itself a never-ending joy. I do not think that doing poor work can ever be a joy; at any rate, not if it is less than one's best, nor can one do one's best at something one does not enjoy. So by all means let us work joyfully.

RANDOLPH WARDELL JOHNSTON

Deerfield, Massachusetts

Preface

To Second Edition

The human spirit has an unconquerable curiosity about life in conditions ordinarily closed to human experience. Hence the fascination of polar exploration, of deep-sea diving, of the conquest of air and space. Hence also, a craft book devoted in large chunks to manipulations of fire in the treatment of metals and ceramic materials. Fire shares this fascination: the ceramist and metallurgist contemplate a world apart, a world inconceivably different from ordinary experience. Metals and rocks considered to be of everlasting hardness become mercurial liquids; the liquids of common use vanish into gases. In this fierce aliveness of the furnace and kiln one can seize upon no familiarities—the atmosphere itself is luminous. There are no shadows, color disappears, or rather all colors are fused into one, the color of heat. All is changed; there is left only the sense of a strange, intense, and totally incomprehensible life, an infinitesimal inkling of the energy of the sun and stars.

A large part of my own life has been spent one way and another in literally playing with fire: fire has been inextricably mingled with my creative efforts in the plastic arts, with my efforts to earn a living, and with the fantasies that have led to technical inventions and innovations.

Preface

There was, it is true, an interlude when the watery element replaced the fiery one in my life. In 1951 we (my wife Margot and our three sons, Bill, Denny, and Pete, then small boys) fitted out a schooner and began a cruise of the Bahamas in search of the ideal desert island homesite. When we had found it, miles by water from any settlement, we set up our first craft workshop in a cave while living in a palm-thatched hut. We were sure that our efforts would be—like Robinson Crusoe's—limited to native materials: as there was no clay, we would work mainly with coconut shells and native woods. But preoccupation with fire seems an ineluctable destiny. With the alternative of much easier ways to earn a living on a desert island, we have, at the cost of incalculable labor, hauling everything in our own boats, chosen to build studios with kilns, melting furnaces, and an electric generating plant. It is clear that one becomes irrevokably commited to a fire-worshipping life. The genes are evidently affected—Pete seems to be going the same way. Now, taking my leave of you as seamen do with their "Fair winds to you!" I wish you "Clear fires and good draft!"

Introduction

Contrary to popular belief the artist or artist-craftsman is the most practical of men, and by managing to find satisfaction or enjoyment in the major portion of his working hours, he strikes a much better bargain with life than his more affluent neighbors whose main interest in work is the effort to reduce the number of hours they spend at it. Hence, although many people prented to look down on the artist for his unworldly attitude toward money, they secretly envy him the privilege of creating, of spending his waking hours at productive and absorbing work, without realizing perhaps that they too could—since all men are artists to some degree—share this satisfaction.

What distinguishes man from animals is—even more than the brain, for other animals have large brains—the hand. And it is undoubtedly working with the hand that has contributed most to the development of what intelligence the human species possesses. An eminent psychiatrist, visiting my studios, once told me that work with the hands alone is properly to be regarded as a hobby from the therapeutic point of view; that he who can hold in his own hands an object made by those hands, achieves an inner poise and balance brought by no other activity. This book then, is dedicated to working with the hands, in particular to working with clay, wood, stone, and metal.

But why this particular choice of materials? Well, they happen to be materials I have worked with a good deal myself. But there is something more: the artist-craftsman is the most practical of men, and he also tends to be the most self-reliant. When he needs a tool or piece of equipment his first instinct is to make it himself. It would be natural for such a man to begin his career in crafts with clay modelling, for which the priceless tools he is born with—the fingers—are all the equipment he needs. As he progresses, however, beyond the use of nature's tools, his first need is for a smaller finger; thus the first wooden modelling tools copy the shape of the finger on a smaller scale. To make these tools the artist must know how to work with wood, as he will also to make the armatures, modelling boards or stands, workbenches, and shelves, with which he will at last equip his studios if he continues to work with clay. A section of this book devoted to working with wood is thus necessary. If the craftsman should become seriously interested in working with wood—carving particularly—he will before long find himself forging or at least modifying and retempering some of his tools; thus we need a section on working with metal. If he should become fascinated with the way sheet pewter and copper can be "formed," that is, actually stretched in some areas and shrunk in others by hammering, he will need to carve hollows in the end grain of logs, and to make wooden mallets. This will send him back to the wood-working section. The clay modeller, meanwhile, has become interested in preserving his models, and needs a kiln to fire them into pottery or terra cotta. This book will show him how to build them. The stone carver, even more than the wood-carver, will need to know how to re-temper his chisels or forge new points or edges on them, so he will need the section on working with metal. If instead of actually carving his designs in stone, he wishes to get in a short time with a minimum of labor the monolithic effect of stone-carving, he can refer to the instructions in cement casting in sand or waste moulds. If he masters the rudiments of sand moulding, it will be no time before he is casting a sand-moulded relief in lead. The appetite for metal casting whetted by this taste, he may feel cramped by the limitations of sand moulding and want to try casting lead in a plaster waste mould that will give him a reproduction in lead of every finger print, every modelling tool mark in the clay. After this experience he will surely be hooked. His life will never be the same again. God knows where he will end—casting in bronze, iron, silver, gold. A new Cellini may be on his way! For he will soon find that if he makes his models in wax, even dancers standing on one toe with extended arms, he can cast them in imperishable bronze in a one-piece mould that, too, will reproduce every personal touch.

Introduction

What seems likely is that many a budding craftsman will become fascinated with the potter's craft, will want to explore hand-building as practiced by the Indians of the Southwest with their manipulations of smoky (or reducing) firing to produce black wares, or oxidizing fires for the red, both from the same clay. He may become intrigued with "throwing on the wheel," which challenging craft demands as much precise balance and physical coordination as water-skiing or tennis.

He who finds himself trapped, possessed, driven by a passion for art, can hardly escape with his life, short of giving that life to art: but as he is unlikely to be free from economic pressures, he must find some way of supporting himself, which usually means some way of reducing the cost of his wares so that they can be sold for a moderate price. One obvious way is to spread the first cost of the original model on which much time has been spent by reproducing it in piece moulds. A limited edition of fifty or less is a good idea. People don't mind paying a little more if the edition is limited. The first set of moulds is likely to be worn out by the time the limit is reached; furthermore, by the time the craftsman has made his limited edition he is fed up with the design and wants to work at something new.

I have not mentioned the thrill of colors matured by the fire. Although there may be few who so feel the call of the divine fire as to give their lives to it, he must be singularly obdurate who, when he opens his first kiln and sees the vivid glowing colors of the fired glazes now exposed to the light of day, is not breathless with wonder. If he should be drawn inescapably to a lifetime search for beauty, he will find that this never ending search is the one essential ingredient in the good life.

R. W. J.
Little Harbour, Abaco, Bahamas

CONTENTS

SECTION I: On Working with Wood

SECTION II: On Working with Clay

Contents

THE BOOK OF COUNTRY CRAFTS

Section I

ON WORKING WITH WOOD

CHAPTER I

Evolution of the New England House; Restoring Old Houses

No one, except a ten year old of Boy Scout urge, is going to build or would want to live in the genuine early New England house. That dwelling was a wigwam, copied from the round-topped elliptical structures put up by the first Americans, the Amerindians.

There is no foundation in the belief that our forefathers who settled first on the New England coast built log cabins resembling the shelters of later frontiersmen. They did not know how. The first American colonists, dumped here by English joint-stock companies, were for the most part an incompetent, helpless and often lazy lot of fellows; in the colonies of Virginia many of them starved to death because they lacked the primitive knowledge of how to shoot a hare or deer, or catch a fish in the abundant streams around. The New England fathers did not fare much better. They came quite unprepared (with the exception of prayer books) for the rigors of pioneering and wresting a living from nature as they found it. It was well over a century after Columbus first sighted this continent before the English were able to set up and keep going a successful and permanent settlement in America.

To expect from these earliest settlers a contribution to architecture is to expect the impossible. Few, if any, had known much beyond the modest cottage. They, like immigrants to other parts of the country, did the most natural thing under the circumstances: they attempted to reproduce as best they could the type of dwelling they had known at home.

Their first experiment, "wattle and daub," an ancient method of weaving boughs and daubing with clay, need not concern us here. Such edifices were soon reduced to mud in our climate. The half-timber house, which is the style everyone terms "old English" and which has been imitated *ad nauseam* by the *entrepreneurs* who construct and sell for profit the awful monstrosities in Long Island and other dark realms of modern suburbia, was also not adapted to the extremes of weather on the eastern coast. After a few trials it was soon abandoned in favor of the clapboard. This last type of dwelling, framed of sturdy, hand-hewn timbers joined without nails, faced with hand-split clapboards, and roofed by hand-split shingles, is the house we can refer to as "colonial." It was a functional structure in the best sense. Its slant-roof helped the snows slide off. Its center chimney, with openings at the back and on two sides for fireplaces, concentrated the heating plant in one up-and-down center unit, just as will be done in the future pre-fabricated structure. Its low-ceiling rooms were easy to heat, and pleasant and cozy in inclement weather. It was convenient to build, and adapted to the country. Its beauty lay in its simplicity and pure functionalism. The same style was followed when the construction was of brick.

As the Americans grew more prosperous—which they did in New England by trading in slaves, rum and molasses, and in ice to the China seas—they began to get more formal. They built larger and more imposing edifices, modeled on the Georgian houses the gentlemen of England inhabited. In New England, whether such houses were found facing a tree-lined common, or in the cities, or even isolated as a single dwelling in the country, these formal structures were known as "town houses," and the financial, moral and social status of a man was indicated by the façade of his house.

The houses we shall have concern with in this book; the houses that people are buying in the country and hoping to fix up, restore and live in, are known under a variety of names: such as the Cape Cod cottage, the salt box, the Colonial, the early American, the pre-Revolutionary, and—later—the Federal, the Greek Revival and the Regency.

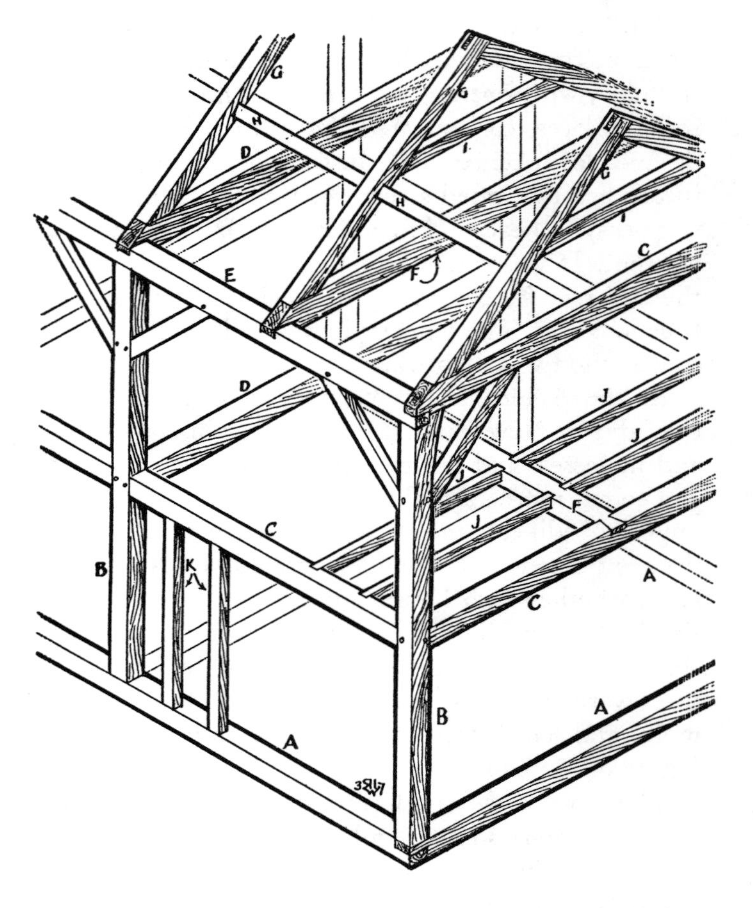

FIG. 1. PARTS OF A TYPICAL OLD HOUSE FRAME: (*A*) *sill;* (*B*) *post;* (*C*) *girt;* (*D*) *chimney girt;* (*E*) *plate;* (*F*) *summer;* (*G*) *principal rafter;* (*H*) *purlin;* (*I*) *collar;* (*J*) *joist;* (*K*) *stud.*

Restoring Old Houses

What should one do with an old house? The question is asked in all seriousness. The site is attractive, the proportions good, the historic associations interesting; but whether to buy it or leave it, whether to preserve it or wreck it, depends precisely upon just what can be done with it. Allow me to suggest that as a starting point our attitude toward the house itself, its materials and workmanship should be one of respect and deference, if not of reverence. There is no reason why our respect for age in human beings should not extend also to age in the product of human hands, human yearnings and love. But there is something else as well. These houses are the intimate products of human effort, the expression of a quality of faith and dignity difficult to find in modern life. There is about them an air of confident assurance, strikingly different from the apologetic lack of conviction which characterizes the imitative eclecticism of contemporary suburban developments, with here a little Tudor, there a touch of half-timber, and yon a sniff of Mediterranean Spanish.

Examining the House

Let us go down to the cellar. I have often questioned the wisdom of having cellars at all, for they are always damp even when they have tight walls and a cement floor. But there is no doubt about the value of that cool, damp darkness as a place to store vegetables; and if the cellar has outside access and outside ventilation, perhaps the dampness can do no great harm to humans.

But it can harm timber, and that is what we came down to look at. You should have your flashlight, penknife and spirit level. So—you go poking about, observing the condition of the timbers, testing them with your knife for soundness, and noting whether they have sagged or whether the cellar walls are falling in. Many of these were "dry walls," i.e., laid up "dry" or without mortar. The term is a misnomer, for, should the house be built on low ground, the water is quite likely to trickle merrily through. Where this has happened often you will find the walls settled or pushed inward and the timbers proportionately out of level. Well, if a timber is much rotted it must be replaced; if sagged much out of level it should be raised.

Examine the frame of the house in the attic and elsewhere to ascertain the condition of the timbers. Renew any of the framing pins which have rotted or become brittle where exposed to dampness.

Reconstruction

If it is bad to put new wine in old bottles, it is equally bad to restore an old house with new materials. There is a shouting disharmony between the impersonal monotony in the surface of machine-made wall-board and the mellowness of genuine hand-hewn beams and hand-planed panels. Our job is to repair the ravages of time in such a way that the house will live again in its original character while adapting it to future occupation. It is not difficult to locate supplies of old timber, boards, brick, and slate, if you have a nose for such things; and they can usually be purchased cheaper than new ones. Hardware stores in country towns always carry old fashioned cut nails, and these make a suitable fastening for old boards, besides being much stronger than wire nails. Also, one can collect old hand-made nails.

Take personal charge of operations unless you have a man who understands the old construction and is not afraid of using old materials. There is a distressing lack of adaptability in many carpenters: they are so accustomed to having everything done for them by machine that they are stumped by a problem calling for a little forethought and ordinary skill with hand tools. They will invariably avoid coming to grips with such a problem by telling you that it is impractical to use old materials, and that it will be much better and cheaper for you to buy new. This is far from the truth. But you need to be on the job to get results.

TO JACK UP SAGGED SILLS

When you jack up a house, you must expect to repair the plaster afterwards. Remove the windows before you start, to avoid breaking the glass. Then, to get the jack under the sill remove some of the stones from the cellar wall. If you are raising a corner use one jack at each side of it. Then screw up gently, watching the effect inside and out, until your corners, sills and eaves are nearest to plumb, level, and true. Then drive in tightly two opposed wedges on a block of suitable height under each timber; nail the wedges securely to the blocks and remove the jacks. Replace the stones in the wall. Mix up some good cement mortar (page 174), and puddle it in with insertions of bits of stone under the timber.

TO REPLACE A ROTTED BEAM

Set up on jacks on suitable solid supports a temporary beam against the joists—parallel, but not too close, to the timber to be replaced. Now screw up your jacks until they just begin to lift the beam. Knock out the old beam with a sledge hammer and crowbar. Set your new beam in place, wedge it up level and puddle cement mortar between it and its permanent support. Be sure that you allow ample time for the mortar to set before removing the temporary beam.

Finish

The finish of the house is an aesthetic as well as a practical problem. The philosophy of these alterations, therefore should be based on the soundest of aesthetic principles: *Eschew all imitations!* An imitation is an ignoble attempt to conceal identity by pretence.

In brief: We shall not

(a) Buy new timbers and hack them up to make them resemble old ones.

(b) Buy cast brass hardware that has been hammered and plated to imitate iron. (This sort of work is a shameful affront to both metals.)

(c) Use cast concrete blocks that imitate the shape of chipped stone, or asbestos shingles moulded to imitate wood, or wood shingles rounded to imitate thatch.

WALLS

The style and period must always be considered. In a country house of the early period, the plaster is somewhat coarse, and the timbers, where they are exposed in the downstairs rooms, hand-hewn. But your plasterer is convinced that nothing equals a perfectly white, perfectly smooth (and perfectly characterless) coating of plaster; and he will be quite insulted when you advise him that the art of plastering is in a period of senile decadence, having long since passed its first vigor. Get him to try out a few samples on a small area of the wall, and stop him before he gets to the putty or white coat, so as to leave the plaster while it still has life in it. Tell him to omit the putty coat and stop at the "scratch coat," but not to scratch it. Use some sand and fiber or hair in the finish coat, and smooth it with a wood float or straight-edge, not with a steel float.

In a town house of the later period you will find

no hand-hewn beams nor rough plaster, but smooth walls and ceilings, and graceful mouldings of delicate proportions. Give your plasterer a free hand. He will take great delight in matching his skill with that of his ancestors.

When you come to buy wall papers, under no circumstances accept a so-called "Colonial Design" that is really the work of contemporary designers attempting to bolster poor work with a dignified name. But a genuine reproduction of an original Colonial pattern may be regarded as merely a new edition of the same thing, when it is printed by a reputable firm. There *are* good original designs of modern wall paper that harmonize with an old house, but they are rare. A safe rule for one inexperienced in these matters, is to use only genuine reproductions of good early designs, put out and guaranteed authentic by a reputable house. Ceilings should never be papered. The manufacture of ceiling papers was an afterthought of the paper-makers when they began to feel the limit of the public's capacity for absorbing their production of wall paper.

A similar case of public taste being turned in a new direction by those who stood to profit by the change, is exhibited in the current craze for "knotty pine." Not many years ago knotty pine was unheard of as a material for anything but cowsheds. But as the lumber companies came to the end of the supplies of good clear timber (a situation that need never have arisen), they turned to the culls of rough, knotty stuff that had never before been considered worth cutting. A cleverly managed campaign of public instruction created the impression that knotty pine was something typically "Colonial," very precious and desirable. Of course, a colonial carpenter having to saw and plane over all those knots by hand would curse violently if he had ever been asked to finish a room with it. Still, it would not be unattractive if it were not finished with that bright yellow, shiny surface and left with the raw newness of its treatment at the mill so obtrusively evident. We should clear our minds, however, of the idea that it represents a colonial custom. If there is one thing against which people should be strongly cautioned, it is against allowing themselves to be stampeded by propaganda from the monthly magazines and selling organizations into an acceptance of any special style or finish as "the" correct and only one. In any case, where you wish to use natural wood, do not finish it with oil, shellac, or varnish alone. (See page 62.)

Sometimes an old house contains pine panels in splendid condition, covered up with layer after layer of old paint. When this is removed, they need nothing more than a patient rubbing with steel wool and paint-remover to acquire a soft, velvety sheen. Do not be disturbed if the color is not perfectly even.

I must caution you against an overzealous exploitation of the "rustic style" in interior finish, exemplified in such acts as building in a house an immense, heavily proportioned cobblestone or fieldstone fireplace, which is suitable only for a hunting camp in the woods. Because of its clumsy scale, such a fireplace seldom harmonizes with the symmetry of the room nor adequately returns to the room the heat from the immense quantities of fuel it consumes. There is also the curious urge to strip lath and plaster from ceilings and joists, and expose, in all their nakedness, the rough-sawn (not hewn) 2″ x 8″ joists. The crowning imbecility in this direction was a ceiling I once saw where between ordinary rough-sawn hemlock joists, carefully divested of their clothing of plaster, and stained brown, had been fitted sheets of white beaver-board. And around the edges were smooth white-painted quarter-round mouldings! Sad to relate, some people have gone to the other extreme and boxed in, with western fir, beautiful, hand-hewn old beams. In replacing these priceless reminders of an age when craftsmanship was man's common heritage, they have achieved the distinction of a ponderous mediocrity: let us hope that they enjoy it.

TRIM

Where additional trim is needed for replacements or alterations it is difficult to match the old material with commercially available mouldings. Find a carpenter who knows how to use old moulding planes and have him make some by hand. If the quantity needed is small this will cost less than having special shapes cut at the mill.

HARDWARE

A number of people have made their own wrought iron hardware by a simple arrangement with the local blacksmith for permission to use his forge and tools. If they get into difficulties, or do not have time to finish everything they start, the master is always there to lend a hand. Or the whole job can be turned over to the blacksmith, who should, however, be encouraged to use his own imagination and inventiveness, rather than asked to follow an exact pattern. Wrought iron hardware should have all the sharp corners and edges filed off and the surfaces rubbed with emery cloth, because it is quite rough when it leaves the anvil. It should be rubbed finally with a little wax.

Hardware should be logically suited to its purpose. A long, heavy spear-pointed hinge applied to a graceful panelled door is an evident misalliance. The rougher, heavier type should be reserved for plank and slab doors, using for the more finished styles either plain butts or the HL type of hinge. Carefully finished brass was often used in the later houses. As there has been no significant change in the method of manufacturing cast brass hardware for generations, good modern brass work should harmonize with an old house, if selected with *taste*. You can take brass hardware as sold, and finish it yourself with files and buffing to a higher degree of finish, if this is desirable, perhaps enriching the surface with a little chasing or engraving. Then lacquer it.

Windows

The very first windows in New England houses were made of oiled parchment or paper, but few houses standing now had such windows in them, for one of the first building materials imported (apart from nails and lime) was diamond glass set in leads, ready to be put in the sash. These small diamond shaped panes were very popular for a while, but soon gave place to larger rectangular ones, set in sash of eight or twelve lights, four across the width and two or three lights high. In some localities the twelve-light sash was below, and in some above the eight-light; and in others a window was made of two twelve-light sashes. The size of the lights was usually 5″ x 7″, 6″ x 8″, or 7″ x 9″.

In the late Victorian period many of these sash were replaced with two-light units of ugly proportions. We need feel no hesitation in again replacing them with eight or twelve-light sash of the original proportions. To keep the delicate proportions the bars or "muntins" should have an outside width of 5/8″ and the sash should be 1″ thick. Stock sash are 1 3/8″.

I believe it is impossible to get stock sash in the exact Colonial proportions. However, special sash ordered in lots of 10 or more are almost as cheap as stock sizes. The nearest I have found in standard sash is the eight-light over twelve "Colonial Windows" of the New York layout. (Standard Lumber sizes are classified either as "New York Layout," used throughout the country, or the "Boston Layout," used chiefly in the vicinity of Boston.) The windows mentioned are made by the Curtis companies and distributed through lumber dealers everywhere. The smallest window fits an epening 3′-0″ x 4′-8″, and has lights about 8″ x 10″.

Outside Finish

Clapboards exposed to the weather for generations without painting acquire a rugged surface and a soft grey or brown tint that harmonizes beautifully with the tones of rocks, lichens and the bark of trees. While it would be a great pity to replace them with new wood, in many cases they are so loose that they require refastening before the house can be occupied. Here is the correct procedure, which, applied to several houses under my observation, has proved entirely satisfactory:

The clapboards are carefully removed and the frame of the house repaired wherever necessary. Then the studs are covered on the outside with a layer of one of the good insulating wallboards, and a layer of tar-paper applied to this again. The clapboards are replaced outside the tar-paper, being fitted as neatly as possible. It is usually necessary to obtain from some other source a few clapboards of a similar age and condition to replace those damaged in the removal. Rock wool insulation is packed between the studs at a suitable opportunity before the job is finished.

If the roof needs renewing you have your choice of two natural materials, wood and slate, and several artificial ones. As in other things, the natural materials inevitably have the advantage of more character. Hand-split pine shingles have a long life and great attractiveness, but are expensive to make. Sawn cedar shingles have the advantage of cheapness but a relatively short life, especially in a damp or shaded location. Slate, an ideal material, is weather-proof, fire-proof, capable of lasting several lifetimes if solidly laid with heavily coated galvanized nails. (It is stupid economy to use uncoated nails for any kind of roof covering.) For early country houses, the rougher, thicker slates are good. Second hand slates or chipped culls are suitable, especially if you can find some with a variety of color. For the more finished style of the "town house" or "Greek Revival," the even, neatly laid regular slates are in keeping. In any case, keep away from fancy patterns of laying. All deviations from the normal weatherage and overlap are traps for the unwary that eventually lead to slates blowing off and to leaks developing.

Artificial roofings lack the character of natural ones, but if they do not pretend to be more than honest roof coverings they can be very satisfactory. An honest roof covering should go with an honest house if the texture and color schemes are in keeping. A dishonest covering, insinuating itself into your notice by its imitation of old natural materials, should be scornfully repudiated. Naturally, no one would consider *roll* roofing as anything but a temporary expedient.

CHAPTER 2

Furniture Construction; Ways of Joining Wood; Furniture Designing

There is incomparable satisfaction in making your own furniture, especially if it is from wood that you have cut down and seasoned yourself, or from old precious pieces of mellow pine, maple or cherry salvaged from some old barn or house. Moreover, if you prize and love these woods, and handle them reverently, you will be already a long way towards giving your work something of that pieceless quality which distinguishes good old furniture—quite apart from its value in the speculative market—from the product of the machine: the quality of being infused with human love and care.

I am going to show the essential features of construction used in nearly all kinds of wood-work, so that you can apply them to whatever you wish to build.

The whole problem of furniture construction may be summed up in one word: *joining*. Since wood does not grow in the shape of tables and chairs with the legs in the right places, we are compelled to de-vise ways of joining one piece of wood to another; unless we adopt the practice of many African tribes of carving their furniture out of solid logs, and even this has its limitations. So, the chief differences in furniture are differences in ways of joining.

A workbench that is solid, strong and true is essential. It should be screwed to the floor and leveled at the same time in each direction with a spirit level. Where practical it is an advantage to have two benches: a rough softwood bench with a strong 2″ top on which blocks and templates of all sorts may be nailed for planing and for wedging up glued boards; and a clean hardwood bench which is to be cared for as a piece of furniture, and on which only finishing and assembling are to be done. You will find it easier to look after tools if you arrange places for them on the wall back of the bench. Everything should be readily accessible—easy to find and easy to put away.

It is also a great convenience to have a zinc or marble covered table for painting, oiling, staining, etc., with a cupboard nearby to contain finishing materials.

Ways of Joining Wood

How to Glue Up a Table Top

True up one face, and the edges to be glued, of each board. Set one board in the vise and try the next one on it to test the joint. If your work is not perfect, the upper board will lean forward or back; or

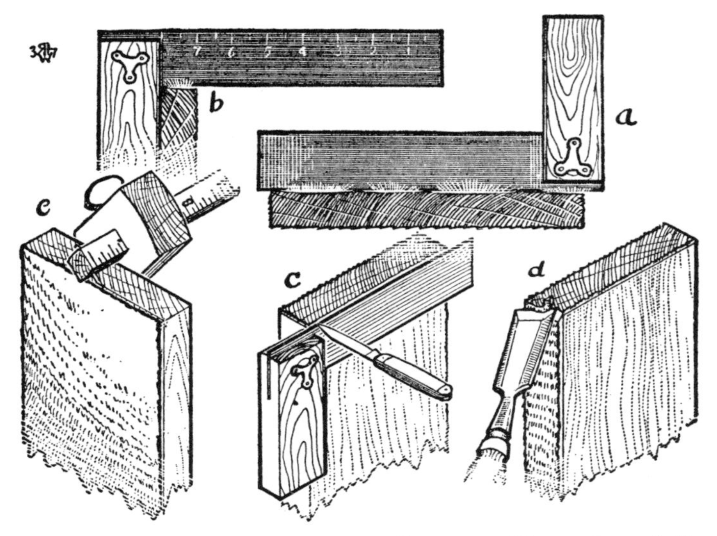

FIG. 2. TRUING UP A ROUGH BOARD BY HAND: (*a and b*) *Testing the work with the try-square by holding it against the light.* (*c*) *Laying out the end.* (*d*) *Preparing the end for planing by trimming the rough corner. You should plane toward this trimmed corner, and finish planing the second end before truing up the second edge.* (*e*) *Laying out the thickness on the edges and ends before planing the second face.*

it may show a good deal of light through the joint. When irregularities are too small to be seen easily, rub some chalk on the truer edge, and then gently replace it on the other. Where the chalk shows, you know that you have to dress off a little of the wood. Repeat this until the joint is perfect. If more than two boards are to be joined, do the same for the other joints. Rub chalk off clean before you glue the joint.

Where great strength is required, and the work is to be exposed to dampness, *spline* jointing is sometimes resorted to. The joint should first be fitted as given above and then the rabbets run in with a rabbeting plane. This has a guide on the side and an adjustable blade in the middle. Different widths of blade can be used; and the blade is adjusted for distance from the side and for depth. The rabbets can also be put in with a small circular saw or a shaper. Dowels (small wooden pins) are sometimes put in glued joints instead of splines to reinforce them. Cut the dowels a trifle short, so they won't strike the bottom before the joint is closed; and try the boards on the dowels before you glue to be sure that they will line up.

GLUES

The craftsman has his choice of three general types of glues.* The oldest is "hot glue", which must be melted over boiling water to use. As it becomes a stiff jelly at room temperature, one must work quickly to brush the glue onto the wood and clamp it up before the glue gets too cold to be squeezed out of the joints. The wood may be heated to prevent this.

To prepare hot glue: buy glue that is hard and clear, with a sharp fracture and no cloudiness. The large sheets have better qualities than the finely ground. Cover it with water and soak over night,

* There are many synthetic glues and adhesives now on the market, including contact cements, and the epoxies, which are extremely strong. Building-supply dealers and hardware stores carry them.

pour off surplus water; and cook in a double-boiler (not enamel) until it thoroughly dissolves and begins to thicken. When a skin forms over it a short time after it has been stirred, it is ready to use.

Liquid glues have chemicals added to keep them liquid when cold. This makes them easier to use than hot glues, but lowers their strength. In damp locations they tend to absorb moisture and become soft. Their one advantage is that they are always ready to use.

Casein glues are made from skimmed milk. The casein is rendered soluble in water by the presence of a strong base, usually ammonia. The base may darken the wood, making the joint show as a clear line where if animal glue is used the joint is invisible. The only other objection is that it must be mixed as required, and if not promptly used, is wasted. It is easy to apply and use, makes strong, waterproof joints, and does not turn the edges of tools as animal glues—hot or liquid—do. It is sold as a powder with full directions for its use.

GLUING THE BOARDS TOGETHER

Two boards can be glued together by a "rubbed joint." Brush the glue quickly onto both edges. Set one board in the vise and rub the other back and forth on it with the grain until so much glue is squeezed out that it will hardly move. Then stop, leaving it in the right position, flush at the edges and ends.

The handiest and safest way of gluing is with screw

clamps. Be sure to place a scrap of wood between the metal ends of your clamp and the edges of your boards to protect them. You can also lay the boards on the floor or on a rough bench, and nail blocks a little further apart than the combined width of the boards, driving the joints up tight with wedges. The boards in either case must be prevented from buckling sideways.

Mortice and Tenon Joints

Use the try-square and double marking gauge to lay out the mortice and tenon, and cut the tenon out with a keen tenon or back saw. Bore holes inside the lines of the mortice and clean it out accurately with a sharp chisel. Secure the joint with dowels through both pieces. The joints should be glued as well as dowelled or wedged. Of the other variations of the joint, the strongest is shown at (h) and (i) in the drawing.

Housed, Dado, Halved and Cross-lap Joints

These are cut out with the backsaw and cleared out with a chisel. Make a level cut by avoiding the tendency to rock the saw.

Dovetail Joints

These are of many varieties. The through dovetail is the simplest and you should master it first.

Lay it out with a sharp pencil and cut carefully just to the lines with a fine tenon saw or a hack saw.

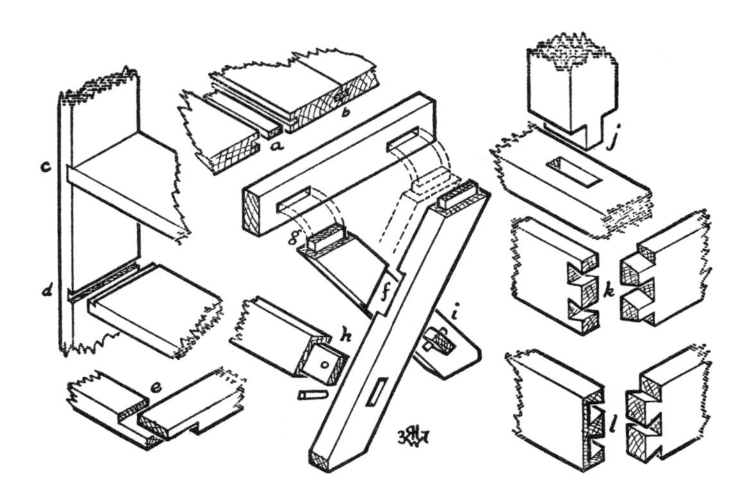

FIG. 3. Typical Joints for Wood-Working: (*a*) *Spline joint, showing spline and rabbets.* (*b*) *Spline joint glued up.* (*c*) *Housed dado joint.* (*d*) *Dovetail dado joint.* (*e*) *Halved or lap joint.* (*f*) *Halved cross lap joint.* (*g*) *Blind mortice and tenon joint.* (*h*) *Through mortice and tenon, with wedged end.* (*i*) *The same, in place.* (*j*) *Stub mortice and tenon joint.* (*k*) *Through dovetail joint.* (*l*) *Lap or half-blind dovetail joint. The centre portion of the above figure shows all the joints used in building an X-end table.*

Then with a sharp chisel and a mallet cut out the bottom of the spaces from both sides of the board. The joint may be glued, but should be tight enough to hold well without glue.

Examine the front corners of old drawers for models of well thought out dovetails.

Mitreing: Picture Frames

It is possible that you share a prevalent impression that in order to make good mitred corners it is necessary to have an expensive metal mitre box and a still more expensive nailing and gluing frame. I offer you the comforting assurance that the possession of these tools does not confer also the power of doing good work if you are not first of all an able craftsman. And if you are an able craftsman you can do perfect work with simple home-made appliances.

MITRE-BOX

The mitre-box shown is familiar to most workers in wood. The chief requirement is to see that, in laying it out, you make the cuts exactly 45° to the sides and perfectly square with the bottom. To get the 45° you need only remember that 45° is the diagonal of a square. Lay out a square across the top of the box by measuring along the side of the box a distance equal to its diameter. Then square across from each end of the line and draw your diagonals. While any crosscut saw can be used, you will find it easier to cut clean mitres with a fine-toothed back saw or tenon saw. You will also find it worthwhile to renew the box from time to time, or at least to make fresh cuts in it when the old ones become worn.

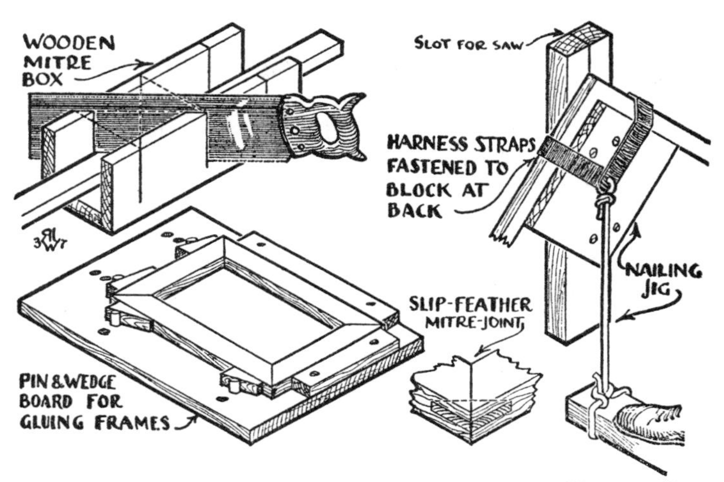

FIG. 4. Mitre Box, Pin and Wedge Board and Nailing Jig.

FASTENING THE MITRES

Small picture frames are usually glued and nailed. Trays, mirror frames and larger picture frames have either a slip-feather joint or a splined joint.

The pin and wedge board shown in the drawing is self explanatory. It is particularly good for gluing up frames for trays where any warping would be impossible to correct after the joints are dry. Slip feathers may be put in by sawing a slot across each corner after the frame is glued and wedged up on the board, inserting a tightly-fitting slip of hard wood with glue, and later dressing it down flush.

The nailing-jig is particularly useful where you have a number of frames to be put together in a short time. Having first cut all your pieces to the correct length in the mitre-box, you begin by inserting two pieces under the harness straps and bringing them together in the correct relationship at the top. Now put your foot on the 2″ x 4″ at the bottom and tighten up the straps. If the joint is not perfect, run your tenon saw through it using the slot as a guide. Then separate the pieces, clean out the sawdust and glue them. When you put your weight on the harness straps again, you may now nail each side of the joint, then step off the 2″ x 4″, move the pieces around and add the next joint. When all the joints are done you slip the frame off the block and start a fresh one.

Steam Bending

The countryman who wishes to make skis, ribs for boats, handles for plows or cultivators, backs, arms or rockers for chairs, will be glad to learn that steam bending is a simple operation that can be carried out with makeshift equipment.

Make a box long enough to contain the wood to be steamed. Leave one end open to slide the wood in. In the middle of the bottom bore a hole to receive a piece of rubber hose; lay wads of burlap on the floor of the box and mount it on trestles, horizontally. Connect the hose to the neck of a five gallon oil can filled with water. Put the can on piled up bricks or a grating, and build a fire under it. Insulate the box with old sacking or quilts, and when it has warmed up, put the wood in. Stuff a wad of sacking in the open end and drape the coverings over it again. The fire should not be too hot—wet steam is better than hot dry steam for this purpose.

The length of time will vary from a few minutes for a piece 3/8″ thick to an hour or more for 1″ lumber. Do not attempt to bend the wood until it is quite soft.

To replace a broken rib in a boat, squeeze the soft, hot, new rib into place, and fasten it at once. For most other work the exact shape should be drawn on a wide pine board, and blocks nailed where needed to hold the steamed wood in place until it is dry.

Furniture Designing

In designing, your job is to keep in mind always the *function* of the thing you are making. Consider that a table, for instance, is a device to bring things within reach of man's relatively short arms. It is, essentially, a board on props to keep it off the floor. These props or legs are so disposed that they do not interfere with the legs of a person who sits at the table. The board and props are securely fitted, so that the table is rigid and strong. If the table is to be moved much, it should be light; if it stays in one place, it may be heavy, if its use so requires. All of these natural limitations of design should be satisfied first.

Apart from them, you can do just what you want, always keeping in mind the definite character of the material you are using. Wood is like nothing else. It is light, porous, warm to the touch, low in compressive strength (easily dented), high in tensile strength with the grain (hard to tear apart), and lower in tensile strength across the grain. It responds to changes in atmospheric moisture by altering its dimensions across the grain but not with the grain,—a property which gives us trouble when we try to oppose the side grain to the long grain: the planks of boats against the ribs, the difference in expansion causing the boats to leak; doors in frames, the tops of the frames being long grain, and the door and its parts side grain; the battens on doors and drawing boards; clapboards on studs, etc.

You can learn a lot from copying exactly a good piece of old furniture. Take a simple piece, from a period that has not reached the full flower of its development. It is useless to think that you can go on from the culminating point of a period and develop further in its style. The artists who went back to the beginnings of periods, and added something of themselves to what they found, built up styles of their own. Those who attempted to copy the *climax* ended in decadence.

If I have not mentioned the style called "modern," it is not because I disapprove of it, but because where it is good, it is so precisely because it is the natural outcome of the mass production methods characteristic of the modern age. But it would be foolish for the craftsman to imitate this product of the machine, as it is for machine-made products to be designed in imitation of hand-made things. Either action is the expression of an apologetic lack of self-assurance.

CHAPTER 3

THE USE OF SMALL MACHINES; WOOD-TURNING

Now that many small wood-working machines are on the market the craftsman will want to know something about their use. In general, machines can be classified into two groups: (a) Those which do things impossible by hand and hence introduce to the craftsman a new art—lathes in particular; (b) Those which merely do things more quickly or easily than the craftsman can by hand.*

Machines of the latter sort have two effects: (1) They make it possible for the craftsman to do more work in a given amount of time, and hence produce more cheaply; (2) They make it possible for him to do tolerably passable work with less skill, and so contribute to the general decline of good craftsmanship. Also, these machines offer serious hazard to

* There are now so many ingenious machines, from power hand-drills and saber saws to motorised wrenches and screw drivers, that one cannot begin to discuss them here. Like cars, models change constantly. You must get the latest information from the manufacturers.

inexperienced (experienced, too) people using them. The accident reports of medical men and insurance companies will show a rapid increase in injuries to hands and fingers from mechanical saws, planers, jointers, and shapers. Yet we are going to see a great deal of these small machines because of their relative cheapness and convenience. It is too bad, in a way. We could stand the increased output of cheap furniture, and the inevitable revival of jig-sawed, gingerbread ornament, as well as the decline of craftsmanship, but it is distressing to think of the beautiful fingers that are to be mangled and amputated.

As the makers of these machines supply full instructions for their use and care, I confine this section to information needed by the craftsman in selecting machines and deciding which he requires for his work.

If craftsmanship is not to suffer extinction from the loss of texture inevitable when hand work is displaced by machine work, we must compensate for the loss by achieving a greater distinction of form. And I mean literally, that we must spend over the drawing board the time we have saved from our hand labor. Our work, having no skill of hand nor richness of texture, will amount to nothing if it has not fine proportion and subtle grace.

Grinders and Grindstones, see page 159.

Circular Saws

The chief advantage of circular saws is that they

cut so finely that, finishing with the plane is unnecessary. Be certain that the spindle runs perfectly true, and that there is not the slightest oscillation of the blade. Spin the shaft by hand, holding a pencil point to the side of the blade to test it. The machine should have an arrangement to tilt the saw to 45°, keeping the table level. It should have adequate provision for guarding the blade; most of the guards are regarded as a nuisance and not used: during such times accidents occur. Circular saws should never be used in classes of young children.

Jig Saws and Band Saws

These are useful for roughing out stuff to be turned on the lathe, and for cutting curved brackets, sections of chair backs and the like; also for ordinary ripping and cross-cutting, although less accurate for this than a circular saw. Jig saws are much less dangerous than either circular or band saws, and will do anything a band saw can do on wood of moderate thickness as well as things a band saw cannot do—such as cutting out the *inside* of pierced patterns on boards. But it is a slower and less sturdy machine than the other. For cutting over 1¾" thick, small jig saws are unsuitable. Do not leave curved surfaces as they come from the saw. Always work them over with the spoke shave, files and sandpaper until the lines of the saw disappear. Neither allow yourself to be seduced by the ease of cutting into a fondness for over-elaborate scroll work. It is unlikely that such can ever be valuable to the design as it is not structurally necessary, adds nothing to the craftsmanship of the piece, and is seldom beautiful in itself.

Belt and Disc Sanders

A combined belt and disc sander is useful for smoothing, especially on rough end-grain and cross-grained wood that is difficult to plane. Since its potential danger is practically limited to "skinning" the fingers or knuckles, it is much preferable to a jointer-planer, for everything except actual jointing. It smooths wide surfaces, and can be used for truing up mitre-cuts, the corners of boxes, drawers, etc.

Jointer-Planers

A jointer that will plane the face of a board up to 6″ wide offers some interest to the craftsman, but since it has no provision for *sizing* boards to a certain thickness, as a real planer has, its value will be chiefly for jointing (truing the edge for gluing). Unless you have a good deal of this to do I should not advise buying one of these dangerous machines. If you are buying finished lumber you can have it jointed at the mill. If you are trimming it with your circular saw it should leave the saw with a perfect edge anyway.

Drill Presses

A drill press approaches the classification of machines that do things impossible to do by hand, since it is almost impossible to drill by hand a hole that is absolutely perpendicular to the surface. It is not a dangerous machine. Doing accurate work, it also saves a great deal of time, especially in drilling iron and steel. It is worthwhile to get a powerful one, otherwise a hand-powered blacksmith's drill is better. It is no trouble to drill ½″ holes through cast iron with a blacksmith's drill.

Shapers

A shaper is useful for running a large quantity of special mouldings. It is simpler and more fun to make small quantities with hand moulding-planes. As shaper knives cannot be reground by the amateur craftsman, he should insist on getting high speed steel rather than carbon steel knives. They stay sharp much longer. The knives should exactly match in thickness and diameter else they cannot be locked to the shaft securely, and are liable to fly out. Some circular saws have provision for use as shapers. As many accidents happen with shapers they should be used with great caution.

Wood-Turning

The lathe is a fascinating instrument that bears the same relation to wood as the potter's wheel does to clay. There is the same magic about it—that feeling of encountering a new dimension. Or perhaps it is the curious feeling you get from working with something that is obviously of three dimensions, yet in which you have command over only two: length, measured on the axis of rotation, and diameter, measured at right-angles to it.

The possessor of a lathe has it in his power to make out of wood all manner of round things; bowls, lamps, chair and table legs, potato mashers, carving mallets, candlesticks, breadboards, cake-trays, napkin rings and tea-pot stands. Wood bowls are ideal for salads, fruits and nuts. Old fashioned wooden trenchers and cups suit the country house perfectly.

The main requirement of a lathe is a shaft that runs true, and a bed that is heavy and solid. Many small lathes are deficient in this respect. The beds are light and the lathes develop vibration, making it difficult to do clean, accurate work. For heavy work of large diameter, investigate the market in used machinery. Heavy, full size lathes can often be purchased for less than the small models. For small work, the bench lathes, scrutinized for points mentioned do very well. See that your lathe has at least three speeds, a full outfit of face plates, tool rests and centers, and chucks for both headstock and tailstock. Mount it securely in a well-lighted situation.

TURNING BETWEEN CENTERS

Long things such as table or chair legs, and the stems of lamps or candlesticks are turned between centers. Cut the wood off longer than your finished work is to be—say about two inches. Find the centre* of each end by drawing the diagonals from corner to corner. Remove the live center from the spindle, set the point of it on the centre of the end of your wood and give it a sharp blow with a wooden mallet (never a steel hammer) until it engages firmly with the wood. Replace the center on or in the spindle, and fit the tailstock to the other end of the wood. The tailstock is clamped to the bed in its approximate position, and the dead center advanced into the centre of the wood. When you have made a fairly deep impression in the wood with the dead center, withdraw it and put a little soap or wax into the space. Then screw the center up and lock it. In doing this get the habit of feeling the headstock pulleys with the left hand while you screw up the tailstock with the right, until you feel a slight resistance to the movement of the spindle. If the resistance is too great the spindle will not turn; if too little the work may vibrate.

Set the edge of the tool rest just below the centre line and close to the wood, without interfering with its free rotation. Form the habit of giving the pulleys a spin with the left hand before starting the machine to be sure that everything is ready.

Start the work on slow speed unless it is small. Use a heavy round-nosed gouge for the first roughing and hold it firmly against the rest with one hand while the other grips it well toward the end of the

* For the sake of clearness cent*re* is used here to represent the *place* on the wood, cent*er* the *part* of the lathe.

handle. In all stages of turning keep a very firm grip on the tool and have it braced steadily. Accidents invariably happen because the worker has lapsed his attention for a moment, relaxing his grip on the tool, gesturing with it toward the work, forgetting that it is revolving at terrific speed, or neglecting to see that everything is right before starting the motor.

Start at one end and pass right along to the other, taking off an even cut all the way. Repeat, readjusting the rest as necessary until the wood is a rough cylinder. Then increase the speed, finish it further with a round-nosed flat chisel. The chisels should be kept extremely sharp with a fine oil stone and leather strop. Poor work will always result if you substitute sandpaper for keen edges.

Set off the longitudinal measurements with a pencil on the smooth surface you have just finished, then hold your pencil still, on the rest, and revolve the spindle by hand to mark the lines all around the wood. Now, turn out the rest of the wood to your lines, following the curves of the drawing from which you are working, using the chisels that seem best adapted to each curve. Use the calipers frequently to check diameters.

If you have a "tapered chuck," a solid socket that fits on the spindle, you may turn the hole in a napkin ring, or the socket in the stem of a candlestick, thus: turn the waste wood on the base of the stem to fit the socket. Remove the work and drive it tightly into the socket. Put the chuck on the spindle, and set a

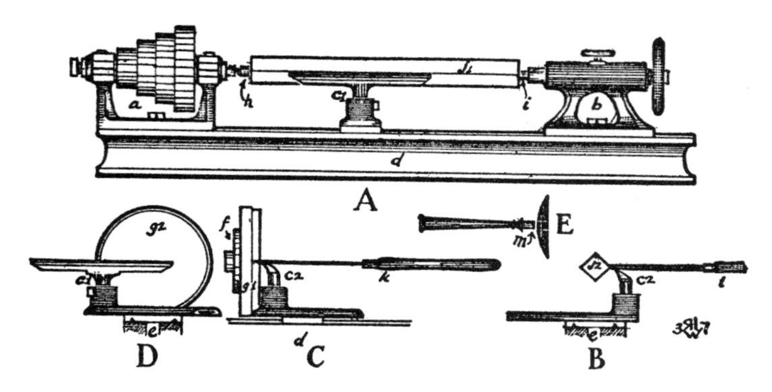

FIG. 5. PARTS OF LATHE: (*a*) *head-stock;* (*b*) *tail-stock;* (*c1*) *front view of tool rest;* (*c2*) *end view of tool rest;* (*d*) *front view of bed;* (*e*) *cross-section of bed;* (*f*) *face-plate;* (*h*) *live center;* (*i*) *dead center.* WOOD-TURNING: (*A*) *front elevation of lathe with square piece of wood* (*j1*) *set for turning between centers;* (*B*) *section of same showing roughing gouge on tool rest against wood* (*j2*); (*C*) *front view showing edge of disc* (*g1*) *being turned on face-plate;* (*D*) *section of same showing face of disc* (*g2*); (*E*) *candlestick turned in two parts showing dowel* (*m*) *turned on the stem and the hole for it dotted through the base.*

lathe bit of the right size (a straight-shank wood bit made for boring on the lathe) into the tailstock chuck. Screw the tailstock chuck up until the hole is cut the required depth.

If you have no tailstock chuck, draw the tailstock away and put the tool rest across the end of your work at right angles to its axis. With a narrow, flat chisel turn the hole out. Then bring the tailstock into position again to steady work for further turning.

The holes for wiring may be bored in lamp stems with a long bit and the tailstock. They may also be bored in the rough wood before turning with an ordinary brace and bit. Bore from each end to meet in the middle. If the bit is not quite long enough burn the wood out with a red hot iron rod. For long stems of floor or table lamps make the stem out of two pieces of wood, rabbetted down the middle and glued together.

TURNING ON A FACE PLATE

Bowls, breadboards, bases of lamps, and candlesticks are turned on faceplates. Plane smooth the side of the wood that is to be the bottom, and on it draw a circle 1/4″ larger in diameter than the finished size calls for. Saw this out and screw to a face plate with short, heavy screws.

Put the face plate on the spindle and set the tool rest across the wood at right angles to the axis. Take a heavy flat chisel and, with slow or medium speed, true up the edge by pressing the chisel against it at right angles to the face of the wood. It is also possible to true the edge from the side position as in turning between centers, but it requires greater care to avoid the danger of splitting off pieces. Next, true up the face of the disc of the wood.

Increase the speed, and shape the wood to follow the drawing, using rule, pencil and calipers. Bore the hole in lamp bases to receive the stem with the tailstock bit, or with a flat chisel. If the hole is deep take care to keep the chisel level and hold firmly. It should be a strong chisel with a long handle. The stem should have its end turned to a kind of dowel to fit this hole. A shoulder is left on the solid part of the stem to support it.

Glue the parts together while the base is still on the lathe. Then turn the dead center up into its place to ensure the perfect alignment of the stem. When the glue is dry touch up the work if necessary, use fine sandpaper, and apply the finishing materials (see page 67) while it revolves at slow speed.

Turn bowls first with the bottom outwards. Finish the outside, mark the bottom with an outline for the face plate. Then put the face plate on the bottom. Turn out the inside and touch up the outside. Bowls, without a "foot" or rim, can be turned directly with the face plate on the bottom from the first.

NOTE: Pieces of large diameter require slower speeds, those of small diameter faster speeds.

CHAPTER 4

Finishing and Refinishing Materials; How to Clean Your Brushes; Wax Polishes; Oil Finishing; Staining and Weathered Effects

Whether you make new things of wood, restore old ones or merely "fix up the house" you will find a knowledge of finishing materials useful. The outline given below emphasizes the effects of old natural wood and warm colors. I shall not discuss paint, enamel or varnish because full directions are printed on their containers. Richness is obtained by bringing out the *natural* qualities of a wood, not by disguising them. The earth colors given below are suitable for tinting wood because they resemble wood colors. Avoid the use of unnatural stains, and do not try to make one wood imitate another.

The beauty of furniture and woodwork is revealed by removing its old paint or varnish and exposing the natural wood. Apply the paint remover with a brush, let it soak a few minutes and scrape with tools that fit the surfaces. Finish with steel wool dipped in remover.

You need not be alarmed on viewing the endless lists of oils, thinners, gums and pigments that are available for treating woods. You can do perfectly good work with quite a limited selection. Some merely replace one another and many are merely

cheaper and inferior substitutes. Here is a list of materials that may be used to make up a simple finisher's kit:

Boiled linseed oil	Dryer
Raw linseed oil	Kerosene
Turpentine	Beeswax
Naphtha or white untreated gasoline	Carnauba Wax
Benzine	Paraffin Wax
Denatured Alcohol	Oxalic Acid
White Shellac	Potash or Soda Lye
Orange Shellac	Whiting
	Silex (Potter's Flint)

Colors

The most generally useful will be ground in oil, but you may occasionally require either dry colors or colors ground in coach Japan. If in doubt or limited to a small outlay get powder colors.

Lamp Black	Raw Umber
Burnt Sienna	Burnt Umber
Yellow Ochre	Other colors as required
Indian Red	

A liberal supply of clean rags is necessary. *NOTE WELL:*—Oily or painty rags are a great source of danger from spontaneous combustion. They should never be allowed to accumulate or even to lie on the bench overnight, but should always be burned or buried or covered with water.

How to Clean Your Brushes

Do not get the habit of standing brushes in water but always clean them out as soon as you stop working with them. If they have been used in oil stains, paints or varnishes, use kerosene or clear gasoline, if spirit stains or shellac use alcohol, if in lacquers use the thinner. This is how you do it: Pour out into an open vessel sufficient of the liquid to dip the brush in. Having dipped it, do not stir it around but quickly withdraw it and hold it upside down a moment so that the liquid may run down into the roots of the bristles. Now put the brush into your piece of rag and wipe it well. You should repeat this until the liquid comes perfectly clean out of your brush, leaving no stain on the cloth. Brushes should be occasionally washed in soapy water. Rub them on the cake of soap working up a lather. Then work them against the side of a basin or on a board, forcing the lather into the roots of the bristles. Repeat until the lather comes out perfectly clean and white, then rinse thoroughly and draw the bristles to the proper shape before setting aside to dry.

Shellac Brushes

It is a great nuisance to clean shellac brushes and if they are not thoroughly cleaned they dry up quite stiff and useless. Moreover, shellac is needed so frequently around the workshop that it is desirable to have a brush that is kept in the bottle. This is very simple to arrange. Take a marmalade bottle holding about a pint and whittle a stopper for it out of a piece of soft pine. See that it is a perfect fit. Then cut out a groove or notch in the side of it that just fits the handle of your brush, keeping the brush up a little off the bottom of the bottle. When using the brush just lay the stopper aside.

Wax Polishes

Wax polishes are useful for protecting surfaces of all kinds. The ingredients are varied according to the kind of exposure the surfaces will receive. Polishes for cars, for example, need to be very hard and bright and so should contain a large proportion of carnauba wax. Wood floors, on the other hand, require a tough elastic finish and the polish should contain a tempering ingredient such as beeswax and possibly a little boiled linseed oil, although the latter will be slower in receiving a lustre. Where a natural wood finish is desired, the wax may be applied directly. For woodwork, paneled walls, and furniture, this preserves the natural color of the wood, if you prefer it to the darker shade brought on by treatment with oil. However, it is not wise to treat floors this way because of the wear they get. To floors apply at least one coat of equal parts of boiled linseed oil and benzine or else a coat of filler.

When the filler has dried, apply a thin coat of shellac and then wax. The filler is made of silex, color, linseed oil and turpentine. Apply it with the grain of the wood, and after it has soaked in rub it off across the grain with rags or waste.

Wax polishes usually have beeswax as a base, and for most purposes there is nothing better. Where you want a harder, more brilliant shine, carnauba wax is introduced, and if you need to lower the cost you substitute a certain amount of paraffin wax. These waxes are melted in a water bath or double boiler and then the pot is *removed from the stove* and the solvent, usually turpentine or benzine, is added. Turpentine makes a tougher film but benzine and gasoline dry more quickly. An example is the following:

Carnauba wax	3 oz.
Beeswax	5 oz.
Turpentine	1 qt.
Benzine	1 pt.

Adjust proportions of thinner, if too soft or stiff when cold.

For applying the wax to floors tie it up in a cloth bag, and squeeze it through the bag as you rub it.

Oil Finishing

For finishing pine and cypress table tops and draining boards for sinks I have found the following treatment unexcelled.

1. Sand perfectly smooth with 0 and 2/0 sandpaper.

2. Apply a liberal coat of raw linseed oil at night.

3. In the morning wipe off any unabsorbed oil.

4. Repeat this every night until no more oil is absorbed. Then continue rubbing with a little oil every day until the surface has a soft sheen all over it.

5. Allow it to harden for several days.

6. Keep it waxed and polished.

Wood finished in this way may be scrubbed with soap and water, and will resist the acids of fruit and vegetables.

Floors may be treated the same way, if the wood is new, clean and free from grease. Another way of treating floors to get a similar effect is to brush the oil on liberally and follow immediately with a vigorous rubbing with a weighted polishing brush. This treatment should be kept up all day long, with one man continually applying the oil and another polishing. Resume work the next day, if necessary, and continue until the floor will absorb no more oil and dries with a soft sheen. It takes a lot of oil, a lot of rubbing, and is rather expensive for large floors. Do not forget to wash the oil out of the weighted brush with gasoline or kerosene.

Linseed oil (raw or boiled) shaken well with shellac in equal parts, makes a good rubbing polish for all kinds of work. It is particularly good for finishing things on the lathe. Be sure that the wood is perfectly finished with the tools and sandpaper first. Apply it with a cloth while the work is revolving at the slowest speed. It will quickly build up a soft polish. For a glassy polish let it dry for a day or so between coats and build up coat after coat until it is as you want it. Or use shellac painted on in successive coats and rubbed down with powdered pumice and linseed oil on a felt rubber, in between coats. If you prefer a soft eggshell finish, substitute water for the oil. But for the very finest high gloss, finish with rottenstone and oil.

Old wood on interior walls takes a soft rich sheen if rubbed with crude oil. The main ingredient here is the rubbing rather than the oil—and this is true of most finishes—there is no substitute for elbow grease.

The finish left by rubbing with paint remover and steel wool is often very satisfactory, there being enough wax in the remover to leave a soft shine if well rubbed.

Staining and Weathered Effects

PIGMENT STAINS

These owe their color to solid pigments mixed with a suitable medium, usually boiled linseed oil, turps or benzine and drier. You can make oil pigment stains by mixing any of the colors given with such a medium using either dry colors or colors ground in oil. The latter will be easier to mix. Shingle stains are made the same way but usually have some creosote for a preservative, and kerosene for a thinner as well as oil. Dark shingle stains may be made by dissolving tar or pitch with kerosene, adding oil, and creosote if desired. Old crankcase oil is even used. If this is allowed to settle it becomes clear and is useful to protect rough wood floors, not to be waxed.

DYE STAINS

These are usually aniline dyes that are applied in various solvents:

Spirit stains: These spirit soluble dyes can be purchased from your wholesale chemist, and dissolved in alcohol to the strength required.

Water stains: These may be purchased as water stains for wood, or purchased as water soluble dyes (such as Diamond Dyes) and mixed to suit your purpose. Alcohol added to the mixture helps it to spread more evenly into the wood. The natural vegetable dyes, particularly the barks, leaves and chips (see page 196) are suitable for making water stains.

Oil stains: Oil soluble dyes are frequently used in oil stains supplemented by pigments.

Chemical stains: In these a chemical combines with substances in the wood, or substances deposited there by another chemical, to form a color right in the wood. Ammonia added to a water stain will give it more "tooth" so it can "bite in." The chemicals given in the section on dyeing fabrics—potassium bichromate, alum, acetic acid, etc. may also be applied to the treatment of wood. *Oxalic acid* is used for bleaching. *Slaked lime* for yellow effects, *potash* for gray, and *potassium permanganate* for old pine. This is a very effective stain. Throw a handful of potassium permanganate crystals into a pail and half fill it with water. Add a pint or so of alcohol, and apply with a brush. Only enough of the crystals will dissolve to keep the solution saturated, and more liquids can be added as it is used up until the crystals are all dissolved. As you apply the stain you can use your judgment in matching the tones—if you need the stain a little darker, dip the brush down nearer to the crystals. When the stain is dry it should be waxed.

Wood treated with water and chemical stains requires sandpapering after it is dry. Stained wood should be protected with wax, oil, or shellac.

CHAPTER 5

On the Art of Cutting, Engraving and Printing Wood and Linoleum Blocks

Wood-cuts, wood engravings and linoleum blocks all have this in common: they are "relief printing," which means they are printed from their highest surfaces, as distinguished from etchings and steel or copper engravings. These are "intaglio printing" or printed from the hollows. In case there is someone who is not clear about the manner in which a wood-block is printed, I suggest the consideration of the rubber stamp, which is surely familiar to everyone.

In a happy marriage there is yielding on one side to balance resistance on the other. So in good printing it is necessary to have yielding coupled with resistance. With the rubber stamp it is the stamp that yields: with letter-press printing it is the paper.

The first wood-blocks were methods of reproducing an outline drawing. This was subsequently developed to include shading, the method always being to draw on the block with ink and have the *formschneider* cut away all the wood around the lines, so that when the block was inked only the lines would print.

All this time with few exceptions the engraver had been an interpreter rather than a creator. One has only to see some of his magnificent interpretations to realize that he did rise to great heights in his field—and one has only to compare his best work with modern half-tone reproductions to see what a characterless substitute we are content to accept in our commerchanical * civilization, where the saving of a few dollars is prized above all else.

A distinct change took place in the art when Thomas Bewick came on the scene in the later part of the 18th century. One of the first to design and cut his own blocks, he is also given credit for first

* *Commerchanical, a word of my own signifying the combination of the worst aspects of commercialism and a mechanistic civilization.*

exploiting the "white line," a perfectly natural innovation for one who was not the interpreter of other men's work. The white line is the mark left by the single stroke of the graver: an artist drawing directly on a block with a graver uses a white line on a black ground. To produce a black line on the block requires an indirect method.

Notwithstanding the impact of Bewick's originality, wood-engraving continued to develop as an interpretive art, becoming ever more minute in its stipplings and hatchings, and tending to compete with copper-engraving, instead of holding to its characteristic boldness.

Arts are subject to the same laws of organic growth as other living things and seldom remain at a level after they have reached the full flower of their development. Thus it seems inevitable that the art of wood-engraving (interpretive) should have decayed and died after reaching the limit of its possibilities, even if the process of photo-engraving had not been invented just at this time.

When the technique of photo-engraving became established as a practical method of reproduction, publishers—continuing to employ artists to make drawings—had the drawings reproduced by the newer, cheaper methods, and discontinued the employment of wood-engravers. Thus with dozens of engravers out of work, and no prospect of re-employment, no apprentices were taken on and the art waned.

Just when everyone thought it was dead for good, wood-engraving suddenly showed signs of life. A few artists found it an ideal medium for a terse, poetic, forceful style of original expression, having little to do with its previous interpretive role. And that brings us to modern wood-engraving.

The great glory and strength of modern work is in the fact that it is a creative art, in many cases cut directly on the block. Now, it is a well known truth about art of all kinds that its expression should be direct, simple and above all not overworked. One achieves freshness by stopping as soon as one has said enough. When one has said enough on a block, only a little of it may be cut away. The resulting proof then has in it much more black than white. This is unfortunate where the design is intended to be printed as an illustration along with type, for the overpowering weight of the black, ill balances with the greyish tone of the letterpress. A thoughtful illustrator will bear this in mind and try to design his blocks so they will balance the type on the same page.

Preparing the Design

The preparations are the same for cutting the three types of blocks: the linoleum-block, the wood-block cut on the plank, and the wood-engraving. Make several alternative studies of your composition with black on white paper, or with white on black paper. I prefer the latter combination, because working with a white line on black is exactly what will be done when the block is cut. These little studies should show only the proportions and the disposition of tones. Select the best design, and where much detail is to be worked out, make a full-size drawing in India ink, the size of the block. Stick it face down carefully with shellac to your block, when it is dry rub the paper off gently with a wet finger, stopping when the lines are exposed.

Or you may first paint your block with India ink, and sketch the design with white water-color. If there is black-line work in it and an excess of white over black, draw on the clean block with pen, brush, and India ink. Whatever the method, the shading strokes should not be drawn, but worked out directly as you cut.

Cutting the Blocks

LINOLEUM

Linoleum is well adapted to large-scale or bold work, but not to fineness of detail; nor does it offer that resistance so challenging in the engraving of box-wood. It may be purchased ready mounted, "type high (11/12"), so that it may be used in a press with type. Or you can glue it yourself onto five-ply veneer and squeeze it in a press over night.

Linoleum is cut with a knife, narrow gouges (veiners), V-chisels (parting tools), and wide gouges for clearing backgrounds. Your problem is simply to

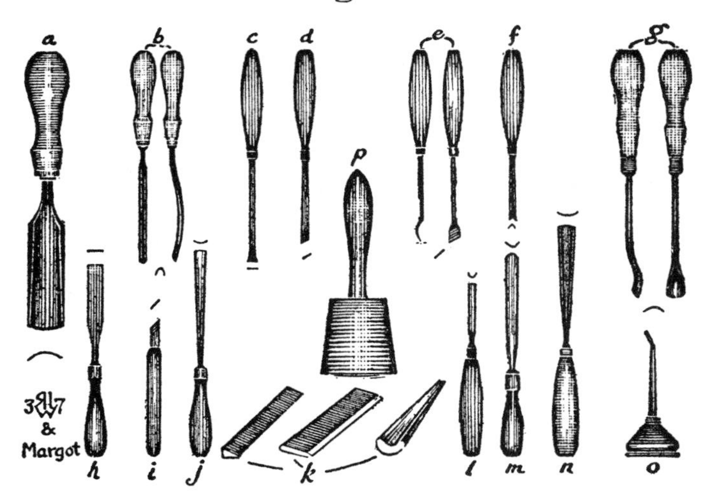

FIG. 6. Tools Generally Used for Wood-Carving: (*a*) *heavy gouge for roughing out large work;* (*b*) *bent gouge for fluting and working in shallow hollows;* (*c*) *spade or fish tail chisel, more useful for carving than the ordinary "firmer" chisel* (*h*)*;* (*d*) *skew or corner chisel, for cleaning out corners, chip carving, and for using as a knife on wood or linoleum blocks;* (*e*) *bent fish tail skew or corner grounder used for levelling off backgrounds;* (*f*) *V-chisel or parting tool;* (*g*) *bent or spoon gouge for deep hollows;* (*i*) *Japanese style of block cutting knife. The best ones are made with a long wood handle which can be whittled away like a pencil as the blade is ground back. It is held somewhat like a pencil;* (*j and n*) *fish tail flat gouges;* (*k and o*) *Arkansas slip stones and oil can, indispensable for keen edges;* (*l*) *veiner;* (*m*) *round nosed gouge;* (*p*) *turned carver's mallet with maple handle and applewood head.*

(*e*), (*i*), (*f*), *and* (*l*) *make a good beginner's set for wood and linoleum blocks. Similar sets with short blades and round handles (like those of gravers) are available. They are useful for linoleum, less so for wood.*

remove to a sufficient depth all the areas you do not wish to print. Large areas must be excavated deeper than the narrow ones; else they will take ink and print. It is easy to see the result of your cutting in white linoleum. The cuts in brown linoleum show best if rubbed with white chalk or zinc oxide. Use the veiners and parting tools as if they were pencils or brushes, lightening dark areas where necessary by hatchings of white lines, and softening the darks into the lights by strokes of the parting tool.

The tool should be held in the palm of the right hand like a graver, the left assisting by pressing back and down on the blade with the first and second fingers. This should be mastered before much cutting is attempted or irreparable damage may result from a sudden slip of the tool. The knife is sometimes used for trimming the edges of dark masses, being easier to use for this than other tools. If you don't possess other tools you can do a whole block with a knife, but it will take longer and give you less freedom of stroke.

WOOD-BLOCKS

Wood-blocks are cut from smooth boards of maple, gumwood, pear, apple, cherry, etc., for finer work; and pine, basswood, poplar, etc., for bolder, open styles. These softer woods will not stand the pressure of the press as well as the hardwoods, but are suitable for printing textiles by hand. For the latter purpose you should use thicker wood, especially if the block is wide, for the wet dye thickenings are likely to warp a thin board. Chestnut is good for this purpose.

Practically all I said about the cutting of linoleum applies also to cutting wood on the plank. You will, however, need to take greater care with wood because of the grain. Keep your tools very sharp. Use a knife and flat chisel more and the veiners less, especially for outlining. Use the wide gouges or spoon chisels to clear away the backgrounds.

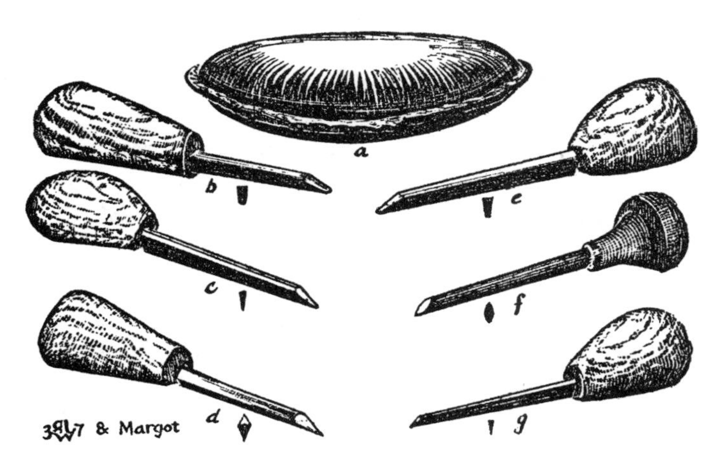

FIG. 7. WOOD ENGRAVING TOOLS: (*a*) *engraver's leather pad filled with sand;* (*b and c*) *round gravers for backgrounds and rounded whites;* (*d*) *lozenge graver* (*the sharp upper edge has been rounded off*); (*e*) *flat graver for clearing rectangular whites;* (*f*) *oval graver or scorper;* (*g*) *tint tool.* NOTE: (*f*) *is shown with a regular turned maple handle, sliced off on the bottom. A much better scheme is to make cork handles, as shown for the other tools, and set the tangs in melted sealing wax.*

WOOD-ENGRAVING

This is done with gravers instead of knives or gouges. The wood is always end-grain. The graver, held in the palm of the hand, is pushed along, taking care to rest the thumb firmly against the edge of the block, to assist in controlling the tool, and to prevent slipping. When working in the middle press firmly downward on the block with the right thumb to steady the hand. If you do not take these precautions the tool is sure to slip and go skating across the block, leaving in its wake a clean white line you can not get rid of.

It is easier to work with an engravers' pad. The left hand holds the block and moves it against the tool, the right hand always remaining in the most natural position for holding and using the tool. Occasionally you may need to work under a glass, but suffer no more of this eyestrain than you have to. Provide support for it at the correct height above your work, and look through it with your best eye, keeping the other open but not focused.

To see the effect rub a little fine white powder (zinc oxide) into the cuts. A better scheme is to match the color of your block with the powder or chalk, so that the new cuts show up in the same value. The matching chalk can be scraped over the block with the bottom edge of the graver every little while.

Proving and Printing the Blocks

Spread a small quantity of printers' ink on a piece

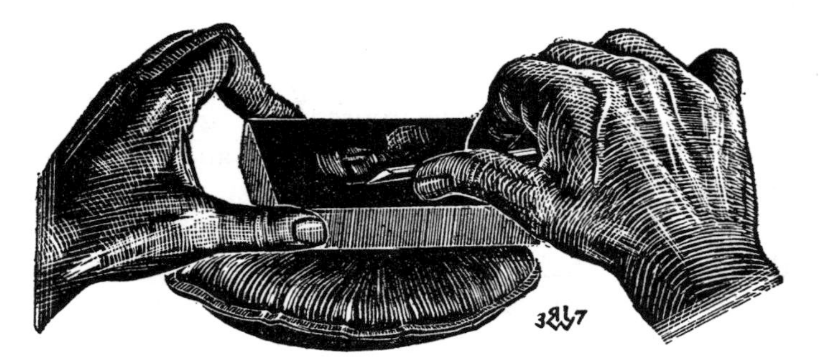

FIG. 8. METHOD OF HOLDING GRAVERS FOR WOOD-ENGRAVING.

of plate glass or polished stone. Then take your dauber or brayer and daub or roll the ink until it is evenly distributed over the face of the slab. Carefully brush all chalk and chips from the block, ink it by rolling the brayer gently back and forth over it, or daubing it until it is evenly covered. On a smooth board place two or three sheets of blotting paper, and on this lay the paper to be printed. Then carefully drop the block on the paper. Slide the board under press until the block is just centered below the screw. Then screw down tight if you are using the old-fashioned letter press. A clothes wringer, or much better, an etching press, or printers' proof press may also be used.

You can also use the burnishing method. Have the inked block face up on the table, lay your paper gently on it and, holding the paper firmly in place with the fingers of the left hand, rub it with the bowl of a silver spoon. Rub strongly from the middle toward the sides, until the design shows through all over. Before you lift the paper off completely, pull up one corner to see if the impression is good. If it is pale and grey in spots, replace it and continue rubbing.

Now look at your proof. Some parts are darker than you expected. Others may be too pale. First note which faults are due to printing and which to the block itself. Also note what work you have to do to complete the design. If you wish to prove it again the same day, don't wash it with gasoline, go on cutting regardless of the ink marks on your fingers. It is easier to see your work this way than if you rub the ink off. If you do, the hollows get stained with ink instead of showing light. Should you need to retouch a stained block, fill up the lines again with the light colored chalk. Prove the block when it is retouched. Try to correct faults in the printing and inking until the impression is perfect. If one end prints darker than the other, put the light end nearer to the screw of the press. If the middle is light and the side dark, pad up the middle of the board on which you print, by pasting small pieces of paper on the board under the light spots, building them up so as to give more pressure to them in the printing. Wash the block, slab, and brayer with gasoline at the day's end, leaving everything clean. Hang the brayer up.

If you are printing on rough hand-made papers it is somethimes necessary to dampen them and pile them between sheets of blotting paper to have them soft enough to take an impression. The Japanese papers are admirable for printing wood-blocks.

If you wish to make your own tools see page 157.

Blocks for Engraving

For printing in a press with type it is simplest to buy the blocks ready-made, especially if they are of boxwood. You can, however, cut them quite neatly with a fine circular saw. See that the timber is dressed quite true on four sides, and set the fence at 11/12″. These blocks should require only a light dressing with fine sandpaper or the scraper. The latter is better, because the sandpaper may leave little bits of grit in the surface and these rub off the fine edge of the graver. However, if you do use sandpaper, take a new sheet No. 1–0 and lay it on your slab, perfectly smooth, and holding the sheet face up, rub the block on it in a steady circular motion. Take great care not to *rock* the block or to let the paper roll up, or the block will not be flat when you are finished.

For woods, boxwood is best, then pear. Rock maple is good for engraving on a bolder scale, but is too soft for fine detail.

When you have your fine blocks printed at last, take care of your prints, and do not exhibit them without the protection of mattes. If the unmounted prints are handled they will soon be spoiled.

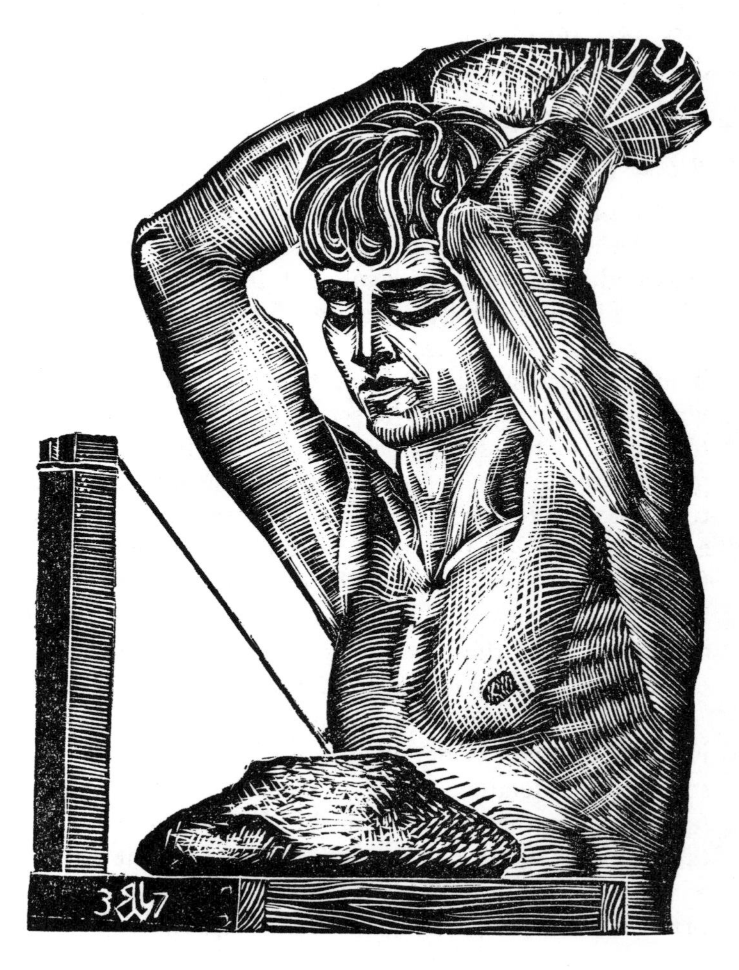

FIG. 9. Wedging Clay.

Section II

ON WORKING WITH CLAY

CHAPTER I

Finding Local Clays; Preparing Clay; Proportioning the Ingredients; Slip-casting

Making pottery is one of the most fascinating pursuits for the countryman. There is adventure in seeking clays; in creating with them, and in developing glazes. There is suspense in the firing; and thrill in opening the kiln afterward; and pride and satisfaction in the useful and beautiful products. Vases, baking dishes, soup bowls, pitchers, plates, beer mugs, tiles, book-ends and candlesticks are among the subjects possible for the amateur. It can become a family project, for pottery making enlists the enthusiasm of young and old alike. I have seen three year olds and grandmothers participating in this art with equal enjoyment. You can buy pottery clays ready for use. But it is fun to seek local clays or to develop a clay body of your own.

In most states of the union a published geological survey is obtainable which offers many hints to the seeker after local clays. Without this, look by the banks of streams, railway or road cuttings, cellar and other excavations. Clay as dug is usually too wet to be plastic when worked up, so a small quantity should be spread out on a board to dry a little, then kneaded to a plastic state. If a strip rolled out the size of a lead pencil, can be coiled around the finger and uncoiled again without cracking, the clay will be found reasonably workable for pottery. Many local clays, however, will not do this and need to be enriched by the additions of ball clay or bentonite.

Preparing Clay

1. Spread the clay out and expose it to the weather. The frost, sun and wind have great value in opening up a clay.

2. Allow it to dry; break it up by pounding and put it through a coarse sieve (¼″ mesh). Take care not to breathe the clay dust, as it is harmful to the lungs.

3. Shake it into hot water, at the same time adding any other necessary ingredients. Let soak over night or longer. Then mix thoroughly to a thick creamy slip. Do not be afraid to put your hands into this slip and feel it for lumps.

4. With a fibre sink brush, rub it through a sieve (16 to 40 wires to the inch, as required) made by

fastening the bronze wire screen, or silk bolting cloth, to a wooden frame.

5. Dry this slip until it is of a plastic consistency, by hanging it in flour or sugar bags, in a warm place open to air currents; leaving it exposed to the air in wooden boxes or tubs; or pouring it into plaster drying basins. The last is the quickest method.

6. Knead the clay until it is soft, smooth and free from lumps. Then "wedge" it: take a fair-sized lump of clay, pass it over a taut wire so that it is cut in half, then dash first one half and then the other forcibly down on the wedging bench. Gather up this mass of clay, knock it into a lump and repeat the process again and again until the clay, being cut through with the wire, shows no air holes. Take care to have your clay sufficiently moist.

7. Put the "wedged" lumps of clay in a damp storage cupboard, made of a box, tub, garbage can or crock. Pile the clay on a board supported on a couple of bricks. Fill water to the tops of the bricks. Then wet heavy cloths and lay them over the clay with the ends hanging into the water. The container should be covered and the water replenished frequently. The clay improves greatly with age; short clays may become quite workable after seasoning two or three months.

The foregoing is the ideal routine in preparing clay. Eliminate one or more of these steps when conditions warrant it. Some clays, for example, are so clean that it is unnecessary to screen them. It is sometimes possible to eliminate wedging, where work is not fired highly, or is unimportant. Where careful work is to be done and much time spent on it, all steps should be followed. Where the firing approaches the vitrifying point of the clay (i.e., the point at which it begins to become dense, non-porous, and glassy) it is absolutely necessary to have the clay perfectly mixed in the slip stage, and perfectly wedged.

Preparing Clays Without Washing

Where it is impracticable or unnecessary to wash clay (i.e., make slip of it and strain it) or where it is desirable to add grog, flint or other grit to ball clays or local clays in the plastic state, then follow this course:

1. Throw the dry lumps of clay into water, hot water if possible. Allow this to soak a day or so and siphon off the clear water. Then let it dry until it is plastic.

2. Weigh out the correct amount of grit to go with the clay, and throw it into water. When it has settled siphon off the clear water and allow it to dry somewhat.

3. Spread part of the damp grit on a clean floor (and see that the whole floor is carefully swept so no dirt will get into the clay) Now lay your plastic clay on the grit about four or six inches thick and spread more grit on your clay. Repeat until the pile is about one and a half feet high.

FIG. 10. Treading Clay.

4. If in winter wear rough high-cut boots, if in summer use bare feet and go ahead and tread the clay down, piling it up again when it spreads out until it is perfectly mixed, without any lumps. Carefully scrape it off the floor, store it away and wash the floor clean again.

This treading of clay should be a social gathering of the young craftsman. There should be music—a fiddle, a guitar or an accordion. There should be two armed with wooden shovels to rush in and pile the clay up when it has been jumped down by the treading partners.

It is best to tread clay on a warm evening in summer. After a day of some gentle occupation such as designing, modeling or painting, after the quiet physical inactivity of the day's work it is good to jump, to dance to music, to feel the blood pounding through one's arteries, to get hot, sweaty, and dirty with clay up to one's knees and elbows. Then to race swiftly across the fields to wash and splash in the warm river under the benevolent radiance of the full moon—then singing on the bank and softly stealing home again.

Proportioning the Ingredients

A craftsman developing a pottery body should have a knowledge of the properties of the various ingredients. They are classified into plastics, including ball-clays and china clays, and non-plastics, including ground flint (more properly ground quartz), ground feldspar, limestone, talc, etc. Drying shrinkage in a body is increased by the plastics and decreased by the non-plastics.

Ball clays approach the theoretical formula for ideal clay Al_2O_3, $2SiO_2$, $2H_2O$, but containing fusible impurities, they vitrify at lower temperatures than china clays. Ball-clays are very plastic.

China clays or kaolins are the nearest to the formula above. They are used to give whiteness to a pottery body and to raise its firing point. Kaolin is the chief plastic ingredient in porcelain, but because its plasticity is much inferior to that of ball-clays, a porcelain body is nearly always short and difficult to handle on the wheel.

Flint (SiO_2) is used to raise the firing point of a body, to open it up where it is too "fat" or "tight," whiten it, and increase its coefficient of expansion. The addition of flint to a body, while actually decreasing its plasticity, may make it easier to work by "opening it up." Where a *glaze* cracks or "crazes" because it shrinks more than the body on which it is applied, the addition of flint to the body will make the *body* shrink more (this refers to the shrinkage that takes place on cooling, not the drying or firing shrinkage), and the glaze will fit it better.

Feldspar (K_2O, Al_2O_3, $6SiO_2$) like flint, opens up a body in the raw clay. It increases the firing shrinkage, because it melts and runs down between the particles of flint and kaolin cementing them together as a glassy flux. This vitrifying effect is complete only at high temperatures.

Whiting ($CaCO_3$) is ground marble or limestone and is frequently used up to about 10% to whiten a body and make it dense. It lowers the firing point of the body.

Talc is used to toughen bodies and lower their firing points. It may tend, however, to shorten the firing range.

Bentonite is a fine colloidal clay which swells up to several times its own volume when added to water. About 7% of the weight of water will make a very stiff jelly. It is useful to add to clays that are "short" to make them more plastic.

Sodium silicate and *sodium carbonate* are added in minute quantities to the water with which slip is made for casting modeled figures and hollow ware in plaster moulds. They make the water suspend a larger proportion of clay than it would otherwise. Although they are sometimes used up to 1/10th of one per cent of the weight of the clay, smaller amounts are usually enough. A dozen drops of waterglass to a gallon of slip is sufficient for most work. If used to excess these deflocculents have just the opposite effect and cause the slip to become very stiff.

In working out your recipe for a clay body, build up the proportions according to the qualities you require. For a starting point you might use something like this, which is a recipe I worked out four years ago and which has proved quite satisfactory in use. Its slight tendency to shortness is largely overcome by seasoning for a few weeks. Bentonite helps this also. The color is a pale buff. May be fired up to about Cone 2 or 3 depending on thickness of pieces. Thin pieces vitrify at this temperature.

Clay Body Recipes

50 lbs. ball clay
10 lbs. kaolin
10 lbs. flint
20 lbs. feldspar
10 lbs. ground limestone or whiting

Another good body, devised to make use of a local blue clay. This body fires a light salmon red, between cones 08 and 03.

50 lbs. ball clay	Sift the bentonite together with the
20 lbs. flint	flint, and add them to the slip after
30 lbs. local blue clay	the clays have been thoroughly
3 lbs. bentonite	broken up into a smooth cream with
100 lbs. water	the water.

If you have no bentonite and wish to use this slip for casting purposes, add two tablespoonfuls of waterglass to the water before soaking the clay in it.

Slip-Casting

Make plaster moulds of the pots or the figures, in two or more pieces, according to the shape. Design the pieces to draw off the figure without pulling the clay. Make the moulds about 2″ thick. Put no waterglass, soap or grease in the mould, and have it clean, and dry. Tie it up and pour the slip into it carefully, to avoid bubbles, keeping the neck of the mould filled over the top for about 10 minutes. Then test the thickness of the cast by scraping against the edge. If it is thick enough, pour out the liquid clay in the middle and leave it propped on battens to drain, until the cast is firm. Removing the mould requires a good deal of care and practice. Note in which direction the mould should be drawn from the cast, and which piece should be removed first, before you tie the mould to fill it.

Touch up the seam after the cast has stiffened up, and if you wish for an even color (almost an impossibility in dark colored clays), model over the whole surface lightly, with a soft cloth.

CHAPTER 2

Simple Methods of Making Pottery

One of the oldest methods of making pottery is with rolls or coils. Roll out between the palms of your hands, a ball of clay about 1/3 the diameter of the bottom of your pot. Lay it on a block of wood or a plaster bat, and gently pat it with the outside edge of the palm of your hand until it spreads into a circle just the thickness and diameter required for the base of your pot. If it grows out too wide, trim with a knife. Practice judging the size of the ball needed for a certain diameter and thickness.

Now gently lift the edges and bend them upward for about 1/2 an inch all around, as a cook might turn up the edges of a pie crust. Roll out a strip of clay about the thickness of your thumb and less than three times as long as the diameter of your pot. This roll should be even in thickness from one end to the other. With light blows of the fingers, slightly flatten it so that it has an elliptical instead of a circular cross section.

Apply one end of this to the *inside* of the turned-up rim of the base (*a*), and with the thumb of your right hand inside, and the index and middle fingers outside, proceed to squeeze it onto the rim of the base, going all around until, returning to the beginning, you trim off the end of the roll and finish it as a complete circle. Then take another flattened roll, apply it to the inside of the previous one and repeat, joining the ends at a different point. You should form the habit of pressing down on the inside of the pot and up on the outside, as you squeeze the rolls together. This spreads the clay from one roll onto that of the adjacent one and broadens the joint between them. You can make a decorative texture, possible to no other type of pottery, by spacing the pinches in an interesting way and not smoothing them off. But if you wish a smooth pot, blend the coils together and fill in any hollows.

When three or four coils have been added, set the pot aside until it has stiffened enough to add more coils. As you proceed you must keep in mind the shape of the finished pot and stretch the rings outward or draw them inward to follow the contour you desire. When the clay is "leather hard," i.e., fairly stiff or about as hard as a milk chocolate bar on a mild spring day, scrape it smooth and thinner. The Indian women use segments of the shells of gourds and calabashes, and pieces of bone or clam shell, ground to different curves to fit the various surfaces of the pots. They are prized greatly, being treasured from year to year. A first-rate tool for scraping the inside of pots can be made from the bowl of a spoon.

You can burnish the clay to a high polish by rubbing with a smooth stone, the back of an old spoon, a tooth-brush handle or a palette knife, especially if you first apply to it a coat of a finer slip.

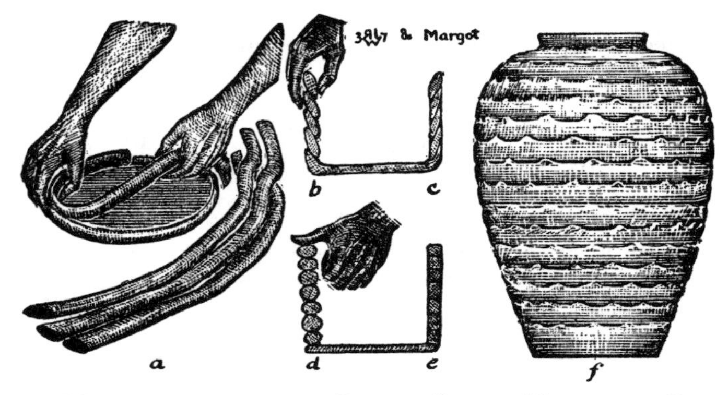

FIG. 11. ADAPTATION OF PUEBLO INDIAN METHOD OF POTTERY MAKING: (*a*) *joining first coil to base;* (*b*) *section showing coils roughly squeezed together. Pressure applied in squeezing the clay firmly together does not distort the walls;* (*c*) *same coils smoothed out, showing long, strong joints;* (*d*) *ordinary method of building as used in most schools, showing difficulty of applying sufficient pressure to join the clay without distorting the walls;* (*e*) *same joints smoothed out, showing short weak joints;* (*f*) *vase in which the squeezing as at* (*b*) *has been allowed to form a decorative pattern.*

There is another method in general use among the Indians of Mexico, for making their everyday cooking utensils, water jars, flower vases and the like. Let me first note a curious fact, that whereas among the Indians of New Mexico, the women make the pots and the men decorate them, among many tribes further south the men make the pots and the women act as their assistants, beating out the clay and also decorating.

Outside the adobe hut the ground is packed into

a hard flat surface. On this the Indian woman beats out the dry clay with a wooden crusher, and here your potter has for a table, a large flat stone covered with sifted ashes. He places in the middle of it a large ball of soft clay, makes the sign of the cross and invoking the aid of the most blessed Virgin of Guadalupe, proceeds to tap it rhythmically with a flat rounded stone dipped frequently in ashes to prevent its sticking to the clay. When the clay is spread to a circular shape, about an inch thick, and fifteen inches across, he picks it up and centers it over the bottom of an inverted "biscuit" (unglazed) pot. Just before this he dashes a little water over the biscuit pot and dusts its damp surface with ashes.

Now observe closely: He pats with the stone dipped in ashes, round and round in even concentric circles, spreading the clay from the middle until it reaches down to the widest part of the pot. Next he takes a piece of clay dipped in ashes and repeats—starting in the very center and tapping in even circles around the pot until he again reaches the widest point. This rounds off the sharp edges left by the tapping of the stone. He dips his hands in water and rubs them over the clay, and finally with a smooth strip of wet linen or chamois he polishes the surface, and finishes by removing the strip from the clay with a dextrous flip, similar to the movement of a boot-black shining shoes. Leaving this to stiffen he applies himself to other tasks. Presently he returns to it, trims it neatly with a knife, and cunningly inverts it

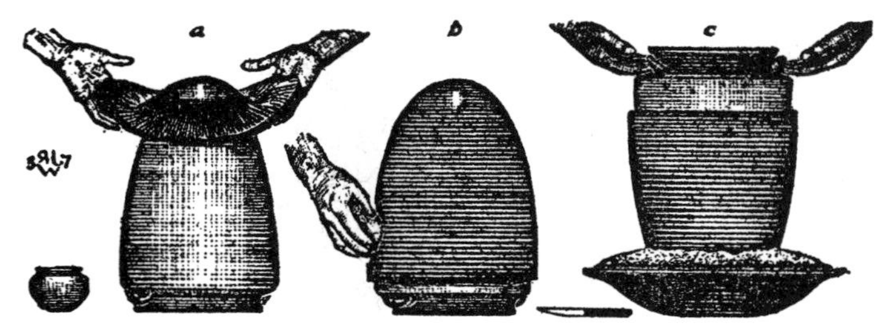

FIG. 12. MEXICAN INDIAN METHOD OF POTTERY MAKING: (*a*) *lifting disc of clay onto pattern, i.e., an inverted pot;* (*b*) *tapping the clay down over the pattern;* (*c*) *withdrawing the pattern from the clay.*

with its biscuit pattern inside it, into a shallow basin filled with wood ashes. He can now lift the pattern out of the clay and finish his pot either by drawing in the top by hand or by attaching with slip another section made in the same manner. Handles or legs are applied while the clay is still quite wet. Should you wish to attempt this method and lack a round-bottomed biscuit pot for a pattern, you may easily find such things as grapefruit, pumpkins and the like, which being covered with a stretched piece of knitted fabric (i.e., old stockings or knitted underwear) will serve the purpose. When removing one of these from your clay, first untie the fabric from it and lift the pattern from the fabric, which in turn will now easily peel out of the clay.

This method is useful to suggest to a group of pupils, asking them to find their own patterns and combine different shapes to make original designs.

CHAPTER 3

Firing Pottery Without a Kiln; How to Build a Simple Wood-Burning Kiln; a Finer Kiln; Oil and Gas Fired Kilns; Gauging Temperature; Setting the Kiln

As a famous authority has remarked: ". . . he who finds it impossible to procure or build a kiln had best take to some other craft." I would only add that if one finds it impossible to procure or build a kiln, one should study and follow the Indian methods of firing without one. Let me tell you of a way I saw the Indian potters of Mexico fire their wares.

They heap broken shards loosely together on the ground and level off the top of the heap. On this they pile the pots carefully together and proceed to build a wall of dried cowdung around them, using pieces of a convenient size. But first they ignite a few pieces and distribute them around the first courses of the wall. When they approach the top they gather the wall in and arch it like a bee hive until the pots are quite covered. The fire glows fiercely from within and white smoke pours from the top. They watch carefully and when the pieces are an even red heat, they knock out the walls of dung and let the pots cool. But if they are firing the black ware, they smother them with loose horse dung, and leave them until they have partly cooled inside the smoky atmosphere. The black color is probably caused by two things—the reduction of the red iron oxide (Fe_2O_3) in the clay to the black (FeO), and the precipitation of carbon in the pores of the clay from the carbon monoxide (CO) in the smoke.

If you follow the Indian method of firing you must also follow the Indian custom of mixing clays with a large proportion of grit to withstand the rapid changes of temperature in the firing and cooling. The types of clay that we fire so gently in our muffles and saggers over a period of many hours would not at all bear the rapid Indian firing which is up, down and over with before two hours are past.

They obtain grit from local sources found good after generations of use. But you, starting afresh, will have to experiment with materials. "Grog" or crushed burned clay—especially crushed firebrick—is ideal. You can buy it; or pound up old firebricks, or unglazed pottery; or crush unburned clay and fire it in other pots. You will find many natural materials that can be used—very fine sand, crushed trap rock, ground up mica, ground soapstone, and especially the flaky sediment from the decomposition of schists and granites, found in small streams. Put it through a thirty-mesh (30 wires to the inch) sieve. If you lack one of these, shake the material into a pail of water stirring it well and pouring the liquid into another pail before any but the largest particles have settled. After a further settlement of a few minutes in the second pail the cloudy water is poured off. The sediment in this pail should be about the right size.

If the clay mixture you get seems to be too coarse, do not forget that the Indians apply a very fine slip to the surface when the pot is almost dry and burnish it by rubbing with small polished stones. You can make this slip by taking the dry scrapings from your pots, mixing them to a thin cream, allowing most of the grit to settle, and decanting the thin slip into another vessel. Then you may add some Sienna or Indian or light red powder if you wish to make it redder, or umber to make it brown. If you wish to make a lighter colored slip for decorating you may add kaolin, flint or whiting to it.

How to Build a Simple Wood-Burning Kiln

The kiln illustrated, fig. 13, is used by the Indians of Mexico for burning their low-fired ware.

The fire is kindled on the ground, and the doorway is kept choked up with fuel, leaving plenty of room inside the fire box and just pushing in the burning sticks from time to time with the fresh ones that replace them. If this is not done, cold air rushes in under the top of the door, chills off the pottery on the near side of the kiln, and short-circuits the draft as well. It is an art to fire a kiln with wood, but it can be learned if one watches not only the fire, but the inside of the kiln as well, observing where the flames play and noting whether any parts of the kiln appear darker than others. If they do, it is a sign that cold air or insufficient heat is reaching the spot and the fire must be manipulated accordingly.

A Finer Kiln

I designed this kiln for burning wood, but it can also be used for coke or coal. The firing technique has been worked out after a good deal of experience with the kiln and should be followed carefully.

The damper (H) is a plate of cast iron, bolted to a rod to give means of controlling it from below with the handle (K). The fire is started with the damper and draft door (D) nearly closed and fuel is added very little at a time. The fire should be very small and should not be built up to its full size for about five hours. Then the draft door and damper are opened and the fuel introduced more frequently. The wood should be kept piled up in front in (B) and fresh pieces are used to push the burning ones off the pile into the fire over the grate-bars (A). Avoid getting too much fuel on the fire itself. No more should be put on it than it can easily consume *at once*, and it should be tended every five minutes toward the end of the firing. Small pieces of wood are better than large. Hardwood sawmill slabs (from the outside of the logs) are the best fuel I have found, especially if they are not too heavy.

Oil and Gas Fired Kilns

These are sold as complete ready-made units and directions are supplied by the makers. They are easier to operate than kilns burning solid fuel, and are always built as *muffle* kilns, the ware being

---

FIG. 13. Plan and Section of a Simple Wood-Burning Kiln Used in Mexico. *It is usually built of stones, but for your convenience in building I have drawn it with a lining of standard firebricks inside a casing of common brick or concrete (such as a concrete culvert). (X) Firebrick set on end flat, against the outside casing. (Z) Arch formed of 12 firebricks neatly chipped and securely wedged as shown. The arch has over it a layer of broken firebrick, flue linings, saggers or other refractory materials to baffle, but not obstruct the flames. After the pottery is set, the kiln is covered (Y) with several layers of old corrugated iron, flattened out cans, etc., loosely laid, to permit the egress of smoke, and supported on iron bars. (M) Spy-hole.*

FIGS. 14, 15, 16. A Finer Kiln Well Adapted for Glazing. *Fig. 14. Plan: sections, above at S′-S′; below at R′-R′. Fig. 15. To the left: front elevation, to the right: section at Q′-Q′. Fig. 16. Vertical longitudinal section at P′-P′.* Parts of Kiln: *(A) grate bars; (B) preheating chamber for fuel; (C) fire door of cinder concrete hung with counter balance, not hinged; (D) draft door; (E) flue, lined with standard 8″ x 8″ lining; (F) double common brick arch to support chimney; (G) firebrick arch made of wedge brick; (H) damper; (I) rod to turn damper; (J) angle irons and tie bolts holding bearings of damper rod and handle; (K) handle; (L) pierced fire arch made of half wedge brick; (M) 6 blocks to raise sagger above fire arch; (N) 15″ diameter round saggers (omitted from Fig. 14); (O) spy-hole; (U) removable door of cinder concrete or Johns Manville "Firecrete."*

Note: *Both kilns must be well bound with iron bands or tie rods at the thrust points of the arches.*

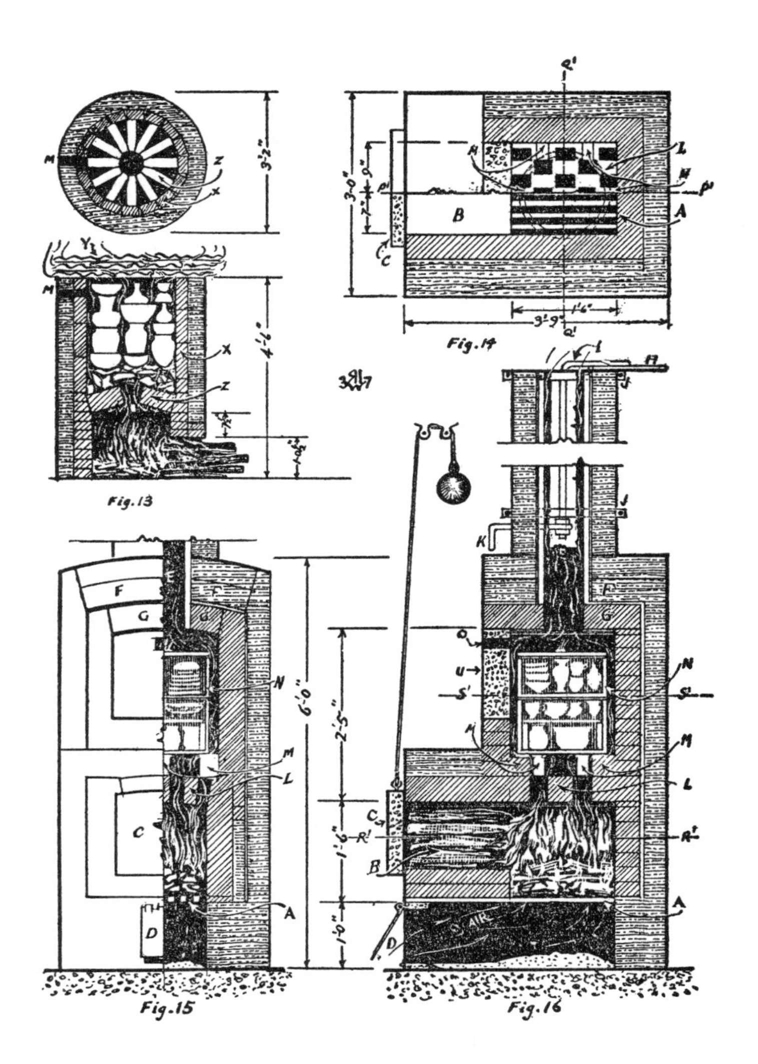

Fig. 13

Fig. 14

Fig. 15

Fig. 16

loaded into the kiln on shelves set on props. Set the kiln up well away from any wood work and as near the chimney as possible. Avoid turns in the pipe connecting with the chimney. Every turn cuts down the effective draft about 25%. A good high chimney is necessary, especially with oil burning kilns that operate by gravity. Take care not to give these kilns too much oil at the beginning of the firing. Apart from the strain on the kiln, of too sudden a heating, you waste the fuel and form deposits of coke in the tubes. Bulletin 133 by Prof. Paul F. Cox of Iowa State College, along with much useful information on clays and glazes, contains plans for a small gas fired kiln that needs no chimney.

A ceramist can build his own oil fired kiln, firing it with a Hauk kerosene torch (operates like a gasoline torch). The torch is mounted so as to throw its flame against baffle plates that support the floor of the kiln. The floor and the baffles are made of 1″ "splits" (firebrick 1″ thick). The products of combustion pass right through spaces in the floor or up the sides, and there is neither muffle nor sagger. This type of kiln is quite efficient and needs little draft, but the terrific noise of the torches takes away from the pleasure of working with it.

Estimating the Temperature

Pottery is fired to alter the chemical composition of the clay by driving off the *combined* water, which ordinary drying does not remove. (Note formulae of kaolin and calcined kaolin on page 111). Once this is driven off, the clay is never softened by water again. The firing also fuses various fluxes and impurities (feldspar, lime, iron oxide, etc.) cementing together the fine particles of clay substance and flint. According to the composition of the body, different degrees of heat are necessary to bring this fusion to the correct point. If carried too far "vitrification" (literally "glassification") results. The firing techniques for porcelain and stoneware are based on bringing the clay body just to the point of complete vitrification and stopping before it starts to sag and melt right down. Pottery is distinguished from porcelain and stoneware by the fact that the firing is stopped at a point short of complete vitrification leaving the body still somewhat porous.

The point at which to stop the firing may be determined empirically by standing up within sight of the spy hole, several pieces of clay with holes in them presented toward the spy hole. At suitable intervals a piece is speared with a sharp poker and brought out. Being allowed to cool quickly, it is examined for hardness by scratching with a knife, tested for porosity by being applied to the tongue (if it sucks strongly to the tongue it is porous and underfired). When the trial piece indicates that the correct point has been reached, allowing for the fact that clay nearer the fire will get a greater heat, the firing is stopped, doors closed and the kiln allowed to cool.

The glaze firing may be gauged by putting a

glazed pot near the spy-hole and watching until the glaze melts. In an open fire kiln, a hole is left in one of the saggers where this method is used.

Most workers use *pyrometric cones.* These are pointed strips of clay substance, so combined with fluxes, or refractory materials, that they soften and bend over at a series of temperatures approximately 30° C. apart.

They are set in pats of clay, three in a row. A softer (or more fusible) cone (for warning), is placed to the right, and a more refractory one (to guard against over-firing), to the left. They are all given an inclination to the right of about 70°. These pats of cones are placed in sight of spy-holes in different parts of the kiln, and the fire manipulated to make them all curve to the same degree at once.

In the small kiln (fig. 16) the pat of cones is placed on top of the sagger cover right in the center. Only one set is used, and the fireman's judgment is exercised in guiding the firing to make the flames play evenly on all sides of the saggers. The spy-hole is placed in the door opposite to this point, covered with a piece of glass or mica, supported by little angles of sheet metal that are set into the concrete or the joints of the bricks.

Setting the Kiln

In setting the kiln consider that the temperature varies with the distance from the fire. The thick pieces need more heat than thin ones, and should be placed accordingly. Modeled heads or figures need a very gentle firing. For most work, fire the biscuit rather soft and then make the glaze firing high enough to mature the glaze and clay at the same time. Plates, however, are better given a high biscuit firing. During their softer glaze firing, they may be supported on stilts, or thimbles, the latter being the more economical of space. Saggers, shelves and kiln furniture of all kinds should be washed with a slip of equal parts kaolin and flint, to prevent glaze from sticking to them. If the glaze is scraped from the bottoms of pots, they may be fired right on the shelves or bottoms of saggers without the use of stilts. Biscuit pieces may be piled together in the kiln, but glazed pieces must always be separated by about 3/8″ to be sure that they do not touch and fuse together.

Saggers or shelves can be made of a mixture of sagger clay or ball clay, kaolin, and crushed firebrick. Saggers and shelves of silicon carbide are best, but have to be made to order by the manufacturer. The kiln in fig. 16 uses two round saggers, 15″ in diameter and 7″ and 9″ high (outside measurements). The bottom sagger is bedded on "wad" made of clay and sand, at six points about its circumference, leaving room between the wads for the upward play of the flames. The upper sagger is covered by a disc made of the sagger-mix. Smaller discs are used inside the saggers for shelves, supported by props of a similar material.

CHAPTER 4

The Potter's Wheel; Building a Wheel; How to Throw

The potter's wheel is the most fascinating machine in the world. The rotary movement it imparts to the clay almost partakes of the character of a spiritual force . . . the two-dimensional, horizontal plane is annihilated or metamorphosed into a single dimension which, returning on itself because of its rotation, has no measurable length and thus may be said to approach infinity. The clay, under this seemingly metaphysical force, grows up as if imbued with life itself, and responds instantly to the merest touch of the fingers. Small wonder that the wheel has been a symbol of creative power for uncounted generations.

If it is true that the master's hands can alter the form of the pot with the slightest touch, it is also true that acquiring mastery in the art of throwing is an arduous pursuit fraught with disappointments and failures. But the reward of effort in this, as in other arts, is in the sense of power which mastery gives; and the accomplishment is well worth all the effort it takes.

Building a Wheel

Most of the ceramic supply houses stock wheels but the price seems out of proportion to the amount of work involved in building them. If you have the money nothing could be simpler than to order one. But do not let the lack stop you, because a wheel can be built for a very little. I remember building my first wheel when I was about twelve years old. I got a blacksmith to make a shaft with a crank in it for which I paid him 15 cents. I set the shaft up in an old table, the bearings being merely holes bored in the wood. I set a wheel of some kind on top and wired a cast iron grate on it for a fly wheel, fastening the head of a paint keg on top of that again for the working surface. The treadle and connecting rod were equally crude. But it worked and gave me much pleasure. I threw pots out of a crude local clay, and fired them by putting them directly on to the red hot coals of the furnace. I cannot now understand why they did not blow to pieces, but they didn't.

The wheel shown (fig. 20) is made of odds and ends. The shaft is an old Ford axle. At top and bottom are the original ball bearings of the car. The flywheel is from an old circular saw and is filled with cement. The wheel head is a clutch plate fitted tightly on the shaft end, and filled with plaster. The plaster was next turned true and recessed to hold bats of a certain size, by nailing a temporary tool rest across and turning the plaster as if it were wood on a lathe. The rounded front guard is half of an old paint pail. And the seat is one of the older style, made of cast iron, from a hay rake. The cast iron ones are much superior to the pressed steel variety. I show this wheel because it has proven very handy. Young children of seven or eight find it easy to use and yet I, who am not a small man, can use it quite comfortably. The built-in seat seems perfectly adapted to the work. The craftsman will find no difficulty in building any one of the three main types of wheels from the drawings given.

FIGS. 17, 18, 19. SIMPLE UNIT FOR VARIABLE SPEED MOTOR-DRIVEN POTTER'S WHEEL. *Fig. 17. Plan, from below. Fig. 18. End elevation of speed-changing device. Fig. 19. Vertical section on line X'X'.* PARTS: *(a) 14" turned cast iron wheel head; (b) turned faceplate with corners rounded, both plates secured to the 1¼" shaft with tapered pins; (c) 1" shaft with long key set in keyway to drive a 4" diameter friction wheel (d). A groove in hub receives notched end of a ¼" x 1¾" forged bar (e), bolted to two ¾" collars (f) which slide on a ¾" cold rolled shaft (g), locked to timbers of table. (e) receives knee of operator, who can shift with it friction wheel from "neutral" (its present position) toward centre of the faceplate. Rectangular opening in table top at (h) is guarded by sheet metal rim (i) to prevent clay or water from reaching bearings.*

FIG. 20. *The kick wheel described in the text.* FIG. 21. *Treadle wheel with a bracket (s) against which to lean.* FIG. 22. *Section of a kick wheel built of stock materials except cast iron head (j) turned to receive bats 6" or 8" diameter, and faceplate (k) turned to fit 1½" cold rolled shaft (l). (n) foot rest, (o) edge of seat, (p) edge of table, all built together; (q) a stock ball-bearing, (r) pipe flange plate screwed to floor with babbit bearing around point of shaft (l).*

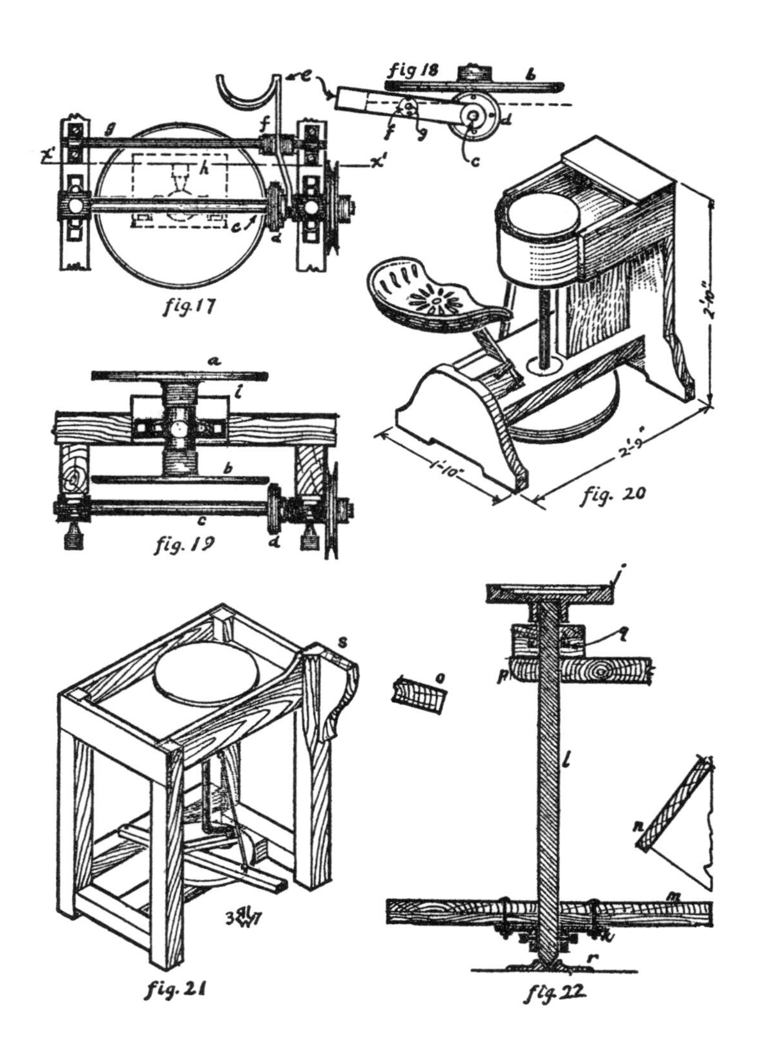

fig. 17

fig. 19

fig. 20

fig. 21

fig. 22

How to Throw

Learn to spin the wheel so that you can do it steadily without conscious effort. See that everything is ready to hand—a few balls of clay as large as good size oranges, a bowl of water, some plaster bats, a piece of wire as fine as thread, and some wooden tools for cleaning up the corners.

The plaster bats are made by pouring plaster into an oiled frying pan, or into plaster moulds made for the purpose. They are discs of plaster about 3/4″ thick and of varying diameters. The wheel head may be turned to receive the bats or you may attach them to it by pressing wads of clay against them all around. A better way is to put a little slip on the wheel head and lay the bat right in the center on the slip. If the bat is dry its absorption of the moisture out of the slip will make it adhere firmly to the wheel head.

The clay should be in good condition, wedged and free from lumps. Set the wheel spinning, moisten the bat slightly and throw a ball of clay smartly onto the middle of it. It will wobble, of course, and your first problem is to get rid of the wobble by "centering." Holding your fingers close together, put the backs of the left fingers against the front of the right, and bend your arms so that the palms come together like a pair of nut-crackers. Practice squeezing with them, using your arms as levers. Then dip your hands in the water, lean forward, keep your elbows tight to your sides, put your "nut-cracker" over the ball of clay and squeeze it while it revolves. If you keep your arms rigidly at your sides, steadying your wrists—if necessary—on the edge of the board in front, the ball will promptly stop wobbling and start to mount in the center. Continue squeezing until the clay has climbed up to about three times its former height; then leave your left hand supporting the base of the clay but apply your right to the top of it and crush it down low again. Clay must always be controlled on both sides; it must be shaped by pressure; so keep a firm grip on it always. Now repeat the squeezing and drive the clay up high again, then crush it down. This should be done three or four times thoroughly to prepare the clay.

When you have mastered this stage, but not before, you may proceed to shape a pot. Take a fresh ball of clay. (Never attempt to use clay a second time on the wheel without first allowing it to dry a little and re-wedging it.) When it is centered and properly prepared by squeezing up and crushing down, put your hands around it in the "nut-cracker" position, and cross the center of it with the thumbs, the right on top of the left. Keeping plenty of water on the clay as a lubricant, spin the wheel and press the thumbs down together. The wall of the pot begins to rise at once. When the thumbs are as far down as they will go, you are to take a new position. Put the hands straight out face to face in the attitude of prayer. Lock the left thumb over the right. Lean forward with the elbows tight to the waist, and turn the hands till they point downward over the right wall of the pot. Now lower them, straddling the wall, the right outside and the left inside of the pot. If the pot is small, you may be able to put only two or three fingers inside it. Press downward with the middle finger of the left hand until you judge that it is deep enough to leave the right thickness of the clay below for the bottom. Then draw it toward the right, pressing the clay between the fingers of the opposing hands so that it rises into the wall. Letting the two thumbs clutch one another firmly, draw the clay upward, increasing the height of the wall until it is of the right thickness.

Practise making cylindrical pots until you have mastered all the stages. You will then find no difficulty in making pots of any shape you wish. If the top is irregular or if you wish to reduce its height, take your small wire wound around a finger of each hand and trim it off, holding one end of the wire inside the pot and the other outside. Use an angular point on a stick to trim out the bottom against the bat. Scrapers or modeling tools of different shapes may be used to thin down the pot and refine its shape while it is still soft. It may also be "turned" with sharp scrapers after it has become "leather hard" or fairly stiff. To do this it is necessary to re-center it upside down on the wheel, and hold it in place with rolls of soft clay. A "foot" or raised edge may be turned at the same time.

FIG. 23.—How to Throw.

CHAPTER 5

ON DECORATING: THREE-DIMENSIONAL; TWO-DIMENSIONAL; COMBINED TECHNIQUES

APART from the essentially decorative qualities of the forms of the pottery itself and the colors of the glazes, decoration of pottery may be classified roughly into two kinds.

THREE-DIMENSIONAL

Three-dimensional is the enrichment by actual manipulation of the surface.

Coil pots easily lend themselves to a decoration of this kind. By rolling the coils carefully and pinching them together at regular intervals patterns can be made of the contrasting texture of the pinched and unpinched parts of the clay.

FIG. 24. POTTERY DECORATIONS. *Modern Mexican Indian (from my own collection)*: (*a*), (*c*), *and* (*k*) PAINTED; (*i*) SCRATCHED *on a black burnished pot, the light body showing through. Modern Pueblo Indian:* (*b*) INCISED *on burnished black pot. This pot was presented to me after I had witnessed its firing, by its maker, "Juanita." Chinese:* (*d*) *and* (*l*). MODELLED. *Early Greek:* (*e*) PAINTED. *Egyptian:* (*f*) PAINTED *border and* (*m*) *pottery hippopotamus; the lotus flowers suggest those of his natural surroundings, the river banks. XVIII century English:* (*g*) PAINTED *on a modelled figure. Early American (Penn)*: (*h*) PAINTED *underglaze on a plate. Navajo* (*j*) PAINTED.

a
b
c
d
e
f
g
h
i
j
k
l
m
& Margot

Lines may be made on a thrown pot while it is revolving, by holding a stick against it.

Decorative forms can be modeled on a damp pot with soft clay, or built up with slip.

Designs may be incised with a modeling tool, cutting out the background and leaving the pattern raised. The reverse is also good.

Two-Dimensional

Two-dimensional is surface enrichment in monochrome or color.

Colored slips may be applied in plain bands or shaded from dark to light, by brushing them onto a pot while it is revolving on the wheel. Underglaze colors mixed with a little clay may be applied to dry or biscuit pots the same way, or used to do freehand brush decoration. This is great fun, and produces charming results. Enough clay should be mixed with the color to insure its holding well under the solvent action of the clear glaze which is to cover it.

Oxides or underglaze colors mixed with clay are pressed into incised decorations and scraped or burnished smooth. This burnished pottery is beautiful if waxed instead of glazed.

Overglaze colors are mixed with a little molasses or gum arabic and painted on top of the glaze, then fired at about cone 015 or lower. This is a very easy way of decorating pottery, but it does not give the depth and character of underglaze work.

Combined Techniques

Model with soft clay or build up with slip of a color different from the background. The French *pate-sur-pate* and the English Wedgewood are well-known examples of these methods, applied to porcelain.

Sgraffito is an Italian version. Dip the damp pot in a slip of a different color. When it is leather hard, cut through the coating of slip, and peel it off in patterns, exposing the body color for a background.

For stimulating suggestions for decorative motifs, I recommend Adolfo Best-Maugard's book: *A Method for Creative Design.*

CHAPTER 6

On Making Glazes; Chemistry of Glazes, Making a Ball-Mill; Applying Glaze; Using Soluble Materials; Overcoming Difficulties; Glazes and Slips for Specific Purposes

The potter should make his own glazes. True, he can buy them ready-made, just as he can also buy the pot ready-made, ready-glazed and ready-fired;

and if all he cares about is to get a perfect pot as soon as possible let him go to the department store and buy one, and forget that ever he wanted to learn the potter's art. Let us never make the mistake of regarding the product before the producer—of regarding what is made more highly than him who makes it—for the great value of any art is not what is made but what happens to him who makes it during its making. For there is no satisfaction like that of mastering a craft from beginning to end, of having command at your finger tips of all materials and processes. All short-cuts and artificial aids are to be put aside as a healthy man would put aside crutches. We should walk every step of the way on our own feet.

Chemistry of Glazes

The modern method of calculating glazes from the "empirical formula" (which might be considered a blue-print for a glaze) replaces the older custom of handing down recipes in potters' families. They did not understand the exact action of each ingredient, nor visualize the molecular patterns, but surrounded the mixing with an aura of mystery close to superstition. Not that they did not produce fine results. For so true is it that pottery is an art rather than a science, that not the whole of our vast army of ceramic technologists, supervising the production of immense factories (a single machine in one factory makes

14,000 dozens of pieces in a day) has produced anything finer than some of the pieces made in small, woodfired kilns hundreds of years ago. The scientific approach, nevertheless, is extremely valuable. It puts into our hands a method of *designing* glazes on paper; of manipulating molecular relationships in a very fascinating way, so that experimenting with glazes is no longer a blind stabbing in the dark, but the projecting of an intellectual concept into physical reality, accompanied by an exciting curiosity as to the outcome.

When you start to design glazes you will need to know something about the characteristics of the various ingredients.

First of all, a glaze is, in effect, a coating of silica or flint (SiO_2) over the pot. Everyone knows that flint is a hard glossy stone, and many people know also that flint and quartz and pure white sand are chemically the same. Flint having an extremely high melting point, we dissolve it with basic oxides usually classified into three groups:

1. The alkalies: Potash (K_2O), Soda (Na_2O).
2. The oxides of the base metals: Lead oxide (PbO), Zinc oxide (ZnO).
3. The alkaline earths: Calcium oxide (lime) (CaO), Barium oxide (BaO).

Alumina (Al_2O_3) or oxide of aluminum is added in relatively small amounts to extend the *firing range:* With it the glaze remains fluid but sufficiently viscous not to run off over a wide range of temperature.

Lead oxide is the most powerful flux and the easiest to use, but the alkalis, potash (K_2O) and soda (Na_2O) are more interesting and also more difficult to use since they are soluble in water. They do not harmfully affect the colors as lead sometimes does. K_2O is brought in by orthoclase or potash feldspar (K_2O, Al_2O_3, $6SiO_2$); otherwise the carbonate or nitrate is usually used. Sodium carbonate or soda ash (Na_2CO_3) is used to bring in Na_2O. This is also brought in by borax. Where borax (Na_2O, $2B_2O_3$) is used, the boric oxide (B_2O_3), being acid, replaces some of the silica and requires a further calculation. Boric oxide has the effect of lowering the melting point and reducing the thermal expansion. You may remember that the 200-inch telescope lens that was cast at Corning, N. Y., a couple of years ago was made of "Borosilicate glass." Boric oxide was used here for these specific properties.

Lime (CaO) is usually added in the form of whiting or ground limestone ($CaCO_3$)—calcium carbonate—a very convenient material because it is not soluble in water.

Zinc oxide is used somewhat to replace lead oxide, also to increase the opacifying effect of tin oxide in opaque glazes and enamels. It is not so strong a flux as lead but is non-poisonous; and having a low coefficient of expansion it is sometimes used to overcome crazing.

INGREDIENTS OF THE CLEAR UNCOLORED GLAZES

Commercial Name	*Chemical Name*	*Formula*	*Equivalent weight*
Borax	Sodium Tetraborate	$Na_2O,2B_2O_3,10H_2O$	For Na_2O content 382 For B_2O_3 content 192
Boric Acid	Boric Acid	$B_2O_3,2H_2O$	124
Flint	Silica	SiO_2	60
Feldspar	Orthoclase	$K_2O,Al_2O_3,6SiO_2$	557
Kaolin or China Clay	Aluminum Silicate (Hydrous)	$Al_2O_3,2SiO_2 2H_2O$	258
Calcined Kaolin	Aluminum Silicate	$Al_2O_3,2SiO_2$	222
Nitre	Potassium Nitrate	KNO_3	101
Soda Ash	Sodium Carbonate	Na_2CO_3	106
White Lead	Basic Lead Carbonate	$Pb(OH)_2,2PbCO_3$	258
Whiting	Calcium Carbonate	$CaCO_3$	100
Zinc Oxide	Zinc Oxide	ZnO	81

Other metallic oxides are used to give color. The typical color is exhibited when it is applied over a pure white body and fired in an "oxidizing atmosphere"—i.e., when the fire has more than sufficient air to burn the fuel with which it is supplied. In a "reducing atmosphere"—when there is insufficient air for the fuel—the carbon monoxide steals oxygen from the metallic oxides in the glaze and "reduces" them, altering their coloring effect:

Name of Oxide or Compound Used	*Equivalent weight*	*Color with Oxidizing Atmosphere*	*Color with Reducing Atmosphere*
Iron (Ferric) oxide, (Fe_2O_3)	160	Reddish brown or yellow	Black
Copper (Cupric) oxide, (CuO)	79	Green	Black metallic lustre, but red with strong reduction
Manganese Carbonate ($MnCO_3$)	115	Brownish purple	Tends to have metallic lustre, and to darken
Cobalt oxide (CoO)	80	Blue	Not very sensitive to reduction
Nickel oxide (NiO)	75	Brownish grey	Not very sensitive to reduction
Uranium oxide (UO_2)	270	Yellow	Tends to turn black
Tin oxide (SnO_2)	150	White	May turn grayish

The composition of a glaze is expressed in chemical terms by the "empirical formula." This indicates the quantities of the different oxides, the sum of the bases always being unity, and the alumina and flint represented by the decimal fractions of this sum. For example:

PbO 0.6			
CaO 0.2	Al_2O_3 .... 0.2	SiO_2 .... 1.6	
K_2O 0.2			
1.0			

These figures represent *numerical* proportions of the molecules of each oxide. Before you could make such a glaze, you would have to work out a "batch" by translating these *numerical* proportions into *weight* proportions of available substances, which would introduce these oxides into the mixture. Set off in a row the fractions and multiply each fraction by the equivalent weight of the substance used to introduce it:

PbO	CaO	K_2O	Al_2O_3	SiO_2		
0.6	0.2	0.2	0.2	1.6		
0.6					White Lead	
0.0	0.2	0.2	0.2	1.6		0.6 × 258 = 154.8
	0.2				Whiting	0.2 × 100 = 20.0
	0.0	0.2	0.2	1.6		
		0.2	0.2	1.2	Feldspar	0.2 × 557 = 111.4
		0.0	0.0	0.4		
				0.4	Flint	0.4 × 60 = 24.0
				0.0		310.2

Notice that the feldspar, calculated for its potash content also brings in sufficient alumina and a large portion of the silica. If the feldspar did not bring in enough alumina, clay (usually kaolin) would bring in the remainder, and it also would bring in silica with it. This addition, or any other, should be set down just in the fashion shown, always referring to the chemical formula of each substance to see what it contains.

Mix a batch weighed out in grams with about a cupful of water, or sufficient to make a thick creamy mixture. While it is best to grind this in a ball-mill, or a mortar and pestle, these operations are not absolutely necessary unless—in the case of colored glazes—one wishes an absolutely even color. I have had good results, with merely giving the mixture a vigorous stirring with a rotary egg-beater.

Making a Ball-Mill

You may, however, easily improvise a ball-mill in this way: make a wooden cover for a stoneware crock and provide it with means for securing the cover firmly to it with a soft rubber washer between. Arrange a frame for the crock that will permit it to be revolved at about 55 r. p. m. for a gallon crock, or a peripheral speed of about 130 feet per minute for any size. Half fill the crock with smooth flint or quartz pebbles about the size of a walnut, and have enough glaze slip just to cover the pebbles. Start your mill turning, and when the time is up (about three-quarters of an hour for a raw glaze), pour out the glaze into a sieve made by fastening 120-mesh bolting cloth, or bronze screen across a wooden frame. Rub it through with a soft brush, rinse the pebbles and mill with water, which is then to be put through the sieve after the first, to wash it out. Then let the glaze settle for a while and syphon or decant the surplus water off the top.

You may also use a wooden keg or churn in the place of the crock. Very little wood is ground off, and it will do no harm to the glaze, since it practically disappears in the intense heat of the firing.

Applying Glaze

The glaze may be applied by dipping the biscuit pot into it, or by pouring the glaze over the pot, by painting it on with a brush—preferably a soft, squirrel or camel hair mop—or by spraying. If spraying is done, have a spray booth with provision for carrying off the exhaust air. Otherwise there is danger to the workers from breathing the suspended lead and silica.

The pot to be glazed must be freshly dusted and quite clean. If the absorption of the biscuit clay is too rapid, dip the pot for an instant in water before applying the glaze. Make the glaze coating as smooth and even as possible; and if you do not make it smooth enough the first time, wash it off in a bowl of water and save the glaze, allowing it to settle as before. Bright (shiny) glazes correct faults in application to some extent, by melting and flowing together in the firing, but matte glazes do not flow at all and show up every defect. The best way of applying bright glaze is to pour the pot full and immediately dump it out, shaking it upside down a moment, and then, holding it by the bottom with the tips of the fingers, to plunge it into the glaze until it comes just to the bottom of the pot, immediately withdrawing it with further shaking. The pot is stood right side up and the fingers marks touched up with the brush. Matte glazes are thickened with gum tragacanth mucilage: approximately one tablespoonful to the pint. (See page 201.)

Here are formulæ and batches you may use as the basis of experiments. When you alter them, work from the formula. Do not, for instance, simply add 100 grams flint to the batch. Keep a clear mental picture of the molecular relationships by increasing the equivalent of SiO_2 in the formula and calculating from that how much additional flint will be required in the batch. This way every alteration will add to your knowledge of the characteristic reaction of the oxides, whereas adding to or taking from the batch is feeling in the dark from which you can profit little.

Bright Raw Glaze, Cone 04-06

PbO	0.80	Al_2O_3	0.2	SiO_2	1.6
ZnO	0.05				
CaO	0.05				
K_2O	0.10				
	1.00				

Batch:				
	White Lead	206	Feldspar	56
	Zinc Oxide	4	Kaolin	26
	Whiting	5	Flint	48

To make opaque add Tin oxide 45 (0.3 equivalent)

Bright Raw Glaze, Cone 1

PbO	0.50	Al_2O_3	0.25	SiO_2	1.75
ZnO	0.10				
CaO	0.15				
K_2O	0.25				
	1.00				

Batch:				
	White Lead	129	Feldspar	139
	Zinc Oxide	8	Flint	15
	Whiting	15		

Matte Glaze, Cone 1

PbO	0.50	Al_2O_3	0.35	SiO_2	1.50
CaO	0.30				
K_2O	0.20				

Batch:				
	White Lead	129	Kaolin	13
	Whiting	30	Calcined Kaolin	22
	Feldspar	111		

Any of the above glazes may be colored with the addition of from one to eight parts of the coloring oxides given in the list. The coloring oxides may be mixed together to obtain different shades but care should be taken not to use too much. The cobalt oxide is the strongest in its effect, one part of it being equal to several of most of the other oxides. Underglaze colors and glaze stains sold by ceramic supply houses may also be ground with the glaze to color it. If you wish to get a mottled effect do not add the coloring oxide until after the batch has been ground. Then merely stir it in. This does not apply equally to all colors, nickel oxide, for instance being too coarse to be added in this fashion.

USING CRUDE MATERIALS

Interesting results are obtained by experimenting with the crude materials: iron rust or scales from the anvil; copper scale from annealing; wood ashes; common blue clay; lead oxide made by melting old pipe, skimming it clean and keeping the metal red hot until it turns to yellowish powder; tin oxide made the same way from old tooth paste and paint tubes, and tinfoil from cheese.

Using Soluble Materials

Soda, borax and potash are sometimes "fritted" (melted with other glaze ingredients) to render them insoluble in water. The mixture is poured into cold water while red hot, to facilitate grinding later.

It is possible to use these substances without fritting by dissolving them in the correct amount of boiling water and thickening the water with laundry starch before adding the non-soluble ingredients of the glaze. It requires experimenting to determine the exact amount of starch and the proper handling. The idea is worth investigating, since it is much easier than fritting and grinding. Also by using the alkalis and borax it is possible to do without the poisonous lead. Not that lead is so very dangerous, but precautions should always be taken; keep cut fingers out of it; do not let it get into the air nor on the floor where it can get stirred into the air by trampling. The lead is not dangerous if fritted first.

This idea might be applied, for a beginning, to the following:

Alkaline Glaze

Na_2O 0.59			
CaO 0.21	Al_2O_3 .... 0.20	SiO_2 .... 1.6
K_2O 0.20			

BATCH:	Grams	For color add:
Sodium Carbonate	62	*For Persian Blue:* 5 gms. black oxide of copper, and 15 gms. tin oxide.
Whiting	21	
Feldspar	111	*Egyptian Blue:* 3 gms. black cop. oxide and 14 gms. tin ox.
Flint	24	

Dissolve the sodium carbonate in one cup of boiling water and stir in one-half teaspoonful of the ordinary laundry mixture of starch and cold water. Then stir the whiting, feldspar and flint into it and continue stirring vigorously. The last three ingredients should have been well mixed with each other by dry grinding, or by shaking them in a box or bottle. No water can be taken away from this type of glaze, as it can from a raw glaze, and your handling technique should be regulated accordingly. ¼ gram of black cobalt oxide added to a batch of the above will make a sapphire blue, and 8 grams of black manganese oxide will make an egg plant color. The colors are sensitive to a great many slight influences and you must not expect perfect results at once nor always, but even the failures are interesting because they are all characterized by the typical softness of texture, which distinguishes an alkaline glaze from any other. The full value of the colors is only developed over a pure white body or engobe.

As a clear glaze it is usually quite successful. At a lower temperature than cone 01 the glaze is less shiny and sometimes looks "frosty."

A HANDY LOW-FIRING GLAZE (*Cone 012 or Lower*)

This is very easy to make and use, and is applicable to low-firing local clays that are used in the crude state, without additions of other material to increase their firing range. It is also good for putting on over underglaze painting.

Na_2O 0.15			
CaO 0.10	Al_2O_3 .... 0.15	SiO_2 .... 2.45	
K_2O 0.15		B_2O_3 .... 0.50	
PbO 0.60			

	Grams
Feldspar	83.5
White Lead	155.
Whiting	10.
Flint	93.
Borax	115.
Boric Acid	25.

Dissolve the borax and boric acid in one cupful of boiling water and add ½ teaspoonful of a mixture of starch and cold water. Stir the other ingredients in and work them together thoroughly with a large paint brush.

The glaze should be applied in about three coats, allowing time to dry in between. Dipping is the best way.

Overcoming Difficulties

You are likely to have difficulties in the application and firing of glazes, and these you will have to learn to deal with. Here are some of them.

CRAWLING: The glaze seems to roll up in mounds or peel right off the body. This is caused sometimes by having an excess of raw kaolin in the mixture. Replace part of it with calcined kaolin (calcine it by putting some finely sifted raw kaolin in a pot and firing it with other pottery in the kiln). The crawling may be caused by dust on the pot before glazing, or a body being fired too high, or by the glaze being ground too fine or being cracked on the pot before firing.

BUBBLING: Where this is a purely local condition affecting only one part of the kiln, it is due to a defect in the firing. All raw glazes bubble during the time that they mature and if the heat is suddenly withdrawn the bubbles harden before they have time to break. A more even firing and slower cooling are indicated here. Over-firing may cause bubbling, the vitrification of the clay closing the pores and forcing air out through the glaze. Where either of these conditions is accompanied by strongly reducing gases in the kiln there is likely to be a dirty, scummy appearance on what should have been a bright glaze. Bubbling or blistering may also be caused by a sulphate scum on the surface of the pot from traces of gypsum in the clay. This may be cured by the addition of about 1% of barium carbonate to the clay body. The affinity of barium for sulphur compounds is well-known and is applied in many industries. Do not use clay so treated in plaster moulds.

CRAZING: Fine cracks spread over the surface of the glaze. This is usually cured by adding more flint to either the clay or the glaze. As this addition to the glaze is liable to raise its maturing point it may then be necessary to further alter the glaze by manipulating the relative proportions of the bases. For this reason many ceramists consider it easier to fit a clay to a glaze than to fit a glaze to a clay.

Glazes and Slips for Specific Purposes

Instead of starting with one formula and making small alterations to it one after another, until you get the desired result, it is simpler to start with two formulæ, each distinctly different from what you want. It is important that they be quite definitely separated in their characteristics. To fit a glaze to a certain body make two batches—one of a glaze too small for the body, and crazed, another of a glaze too large for the body and tending to shiver off. Then make tests on small tiles or fragments of clay with mixtures of these two glazes: 1 part of A to 9 of B, 2 of A to 8 of B, 3 of A to 7 of B, and so on, measuring the glaze slips out with a tablespoon or measuring cup and keeping the stock glazes thoroughly stirred while you work. Out of these tests select one or two and work on them further. Use the same system in working out a clay body. Calculate the correct formula of the best sample from the proportions of the two contributing slips.

CHAPTER 7

How to Make Tiles

What could be more delightful than to have the sides of your fireplace, or the walls of your kitchen, bathroom or playroom decorated with bright tiles made by you or your children? Like ducks to water, children take to designing tiles, and produce lively and engaging decorations.

Where children are making individual tiles let them make balls of clay, and flatten them gently by pounding with the fleshy part of the palms. Each child should have a plaster bat or a dry pine board a little larger than the tile he is making. The tiles are trimmed to the correct shape and size with a knife. More uniform results may be obtained, if you nail to the pine board two parallel strips of wood as thick as the tile and separated by the width of the tile. The clay is pressed down between the strips and rolled with a dry rolling pin. A plaster roller is better, it can be used longer before getting too damp to leave the clay without sticking. The clay is left in the board until it is dry enough to be tipped out.

The designs are drawn on thin paper and traced on the clay. The figures may be left standing up and the background cut away, or they may be merely painted with colored slips or underglaze colors. Or the recessed part of the design may be filled with clay of another color (natural or tinted with oxides), scraped and burnished smooth. This is very suitable for tiles that are to be waxed after firing, instead of glazed. Where a large number of tiles are needed, as for a floor, a wooden mould is made. The mould is wet with water and a handful of fine grog thrown in and dumped out. The clay, being quite soft and having a good proportion of grog in it, is pressed vigorously into the mould and straightened off with a stick. There should be a supply of boards a little larger than the tiles. Place one of these over the mould with the clay in it, and invert the two, so that the tile falls onto the board. Lift the mould off, throw in fresh grog and repeat the operation.

Tiles may be decorated by pressing on them little dies carved out of wood, or made of plaster or terra cotta. Unless they are very dry it is best to lubricate them with oil, talc, or powdered mica. You can fill in the recessed part of the tile with colored or opaque white glazes, and leave the surface of the tile unglazed. Tiles finished in this way are easier to fire.

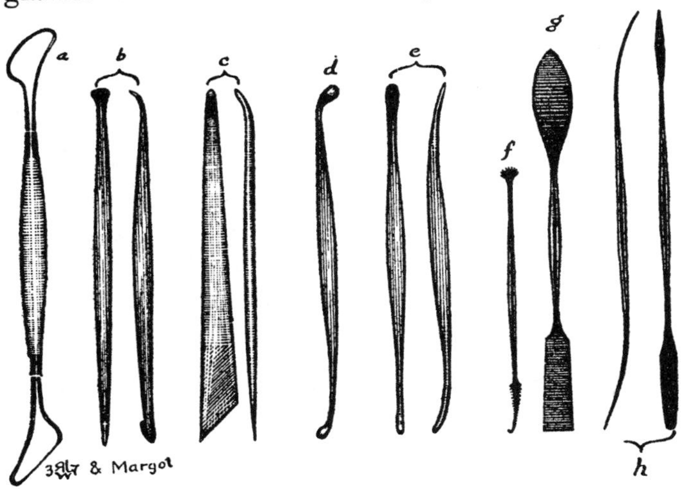

FIG. 25. TYPES OF TOOLS USED FOR WORKING WITH CLAY AND PLASTER. *Make them as follows:* (*a*) *turned maple handle and spring brass wire ends;* (*b*) *to* (*e*) *boxwood or lignum vitæ, rasped and polished to shape;* (*f*) *to* (*h*) *forged cast steel, ground and tempered.*

FIG. 26.—WORKING WITH METAL.

Section III

ON WORKING WITH METALS

CHAPTER 1

How to Work with Sheet Metals; Tin; Pewter; Copper, Brass, Nickel Silver and Sterling Silver; Soft Soldering; Shaping by Hammering in a Hollow Block; How to Shape Cooking Vessels; How to Line Copper Cooking Vessels with Tin

Tin (which is really sheet steel with a thin plating of tin) is the traditional New England material. Itinerant tinsmiths used to travel through the country, peddling needles, thread and buttons as well as doing tinkering and repairing. They made sconces, candlesticks, pie and cake tins, spice cans, etc. Many of these old products of the tinsmith's art are still in the attics of old homesteads. They serve as interesting models for the metal worker to study as starting points for his own designing. A great deal may be learned about the craft by measuring one of these pieces and attempting an exact copy. Paper patterns of each part should be made and fitted before cutting the metal. A knowledge of mechanical drawing, particularly the kind known as "Development of Surfaces" is useful to the worker in sheet metals, especially to the tinsmith who has to work out the patterns for new shapes.

Tin is made in such thin sheets that it is difficult to alter from its characteristic flatness by hammering, as one does with copper and pewter. Also its cheapness is such that the work of hammering is not warranted; since it is an established rule of craftsmanship that the more precious the material, the more work one is justified in spending; the reverse being equally true, that the cheaper or baser the material the less time one should spend on it. Tin must be treated very much like paper. If you wish to design for tin, you have but to work out the pieces in paper first, paste them together, bend them, curl them and twist them as you please. If you allow enough to paste the pieces together, say ¼″ to ⅜″ that will likewise be sufficient for soldering the tin. Cut your patterns double. Then keep the pasted model and use the extra patterns for tracing the tin.

Curved frills radiating fanwise about the borders of candle sconces are put in with the wedge shaped end of a tinsmith's hammer, driving the tin into a carved notch in the end of a hardwood log. Their decorative quality is incidental to the fact that they are necessary to gather up the edges of the tin and draw them forward. Small convex bosses (such as feet for trays or candlesticks) can be driven out with a ball-pein hammer over little depressions in the end of the log.

Pewter

The old pewter, so well-known in New England, displays a rare perfection of finish that is almost beyond the skill of the modern amateur, to whom all of the forms, moulds and special tools as well as the traditional training of the old craftsmen are not accessible. Nevertheless, his aim should be toward the acquiring of something of that skill. I do not mean that he should carry this so far as to produce work that shows no evidence of being made by hand at all; but he should aim at going beyond the ordinary summer camp production of hammered plates and ash-trays. The metal, being fairly costly, deserves to be enriched by having a fair amount of labor applied to it in the working out of more complicated problems, such as pitchers, deep sugar bowls with fitted lids, hot water jugs, teapots, mugs and the like. A little chasing or engraving judiciously applied as a border may add a great deal to a piece that is already very satisfying in form.

We should keep a nice sense of fitness in expending our skill; not attempting to follow out the fineness of the goldsmith's art in dealing with a commoner material, but going rather for a freer, more rapid and exuberant style.

How to Solder Pewter

Fit the pieces to be soldered as closely as you can. Have both sides of the joint perfectly clean. Support the two pieces so that they stay together in the correct position and apply special flux to the joint. Take in your left hand a wire of special pewter solder, and hold in your right a bunsen burner or an alcohol lamp to which is attached a blowpipe. Blow gently and steadily, and focus the point of the flame on the end of the solder wire, which should be applied to the beginning of the joint. Bathe the surrounding metal gently with the flame until the solder melts and runs into the joint, and proceed along until it is finished. Never attempt to solder pewter with a soldering iron or ordinary solder.

Special Flux for Pewter, Tin, Brass, and Copper

	by volume
Glycerine	5 parts
Zinc chloride	3 parts
Distilled water	12 parts

Special Solder for Pewter

	by weight
Tin	2 parts
Lead	2 parts
Bismuth	1 part

Weigh out the metals and melt them in a ladle or small tin can. Bend a pouring lip in the can with the pliers. Then pour the metal into small grooves gouged out of a pine board. Don't forget to stir the metal well, and to skim it before pouring. Zinc chloride is made by dropping zinc into muriatic acid until the effervescing stops.

Copper, Brass, Nickel Silver and Sterling Silver

All of these metals are treated very much alike. They require annealing after they have been hammered for a little while. To anneal these metals make them red hot with a gasoline torch or over a forge, and plunge into cold water. The number of annealings on any piece of work should be as small as possible. Never attempt to anneal aluminum or pewter in this way because they melt at lower temperatures than red heat.

How to Solder or Braze These Metals

All of these metals may be joined with silver solder. Clean the joints thoroughly, coat them with a paste of borax or boric acid and water, support the pieces in the proper relationship and apply the heat. As the metal gets red hot lay strips of the silver solder against it. Or these strips may be set in between the pieces of metal before it is heated. Take care not to heat any part of the metal to a brighter color than the rest. This especially applies to brass, which may contain so much zinc as to make its melting point almost as low as that of the silver solder. When the solder has run into the joint and filled it, allow the work to cool. Copper is usually *brazed* instead of silver soldered, because brazing spelter is much cheaper. The spelter is a low melting brass (about equal parts copper and zinc) used in the form of fine granules or filings. These are mixed with water and borax to a paste and applied to the parts to be joined. Then the work is heated as for silver solder.

Soft Soldering

The ordinary solder of commerce, known as soft solder, is usually made of equal parts tin and lead ("half and half"). It is suitable for joining tin, galvanized iron, zinc, and such articles of copper and brass as do not by their use or value warrant the trouble of hard soldering.

The usual tool is known as a "soldering iron," but it is never made of iron, always of copper. It is heated to about 600° F., which can be judged by holding it about 2″ away from the cheek, at which distance it should feel fairly hot. Before it can be used the iron must be "tinned." The hot iron is rested against a rough piece of wood, and filed bright on one face. Immediately a piece of rosin is applied to the new metal and withdrawn. Quickly solder is rubbed on the point, and it flows promptly over the new surface of the copper. This is repeated for the other three sides of the point. Tinning must be renewed as it gives way to the black scum which forms while the iron is hot, especially if it is accidently heated to redness. Electric soldering irons are useful but do not hold much heat.

"Sweating" soft solder is done without an iron, in the manner described under pewter. Articles of brass, copper and tin may be wired together in the correct position, having first been thoroughly cleaned and fluxed, and heated to about 400 F. The solder wire is fed along the joint until it is complete. To be sure that the solder "takes" all over the metal you may, before sweating the parts together, "tin" all the surfaces to be soldered.

"Tinning" metals is accomplished in the following manner: (1) With steel wool or fine sandpaper rub up a strip on the edge of the piece to be joined. If the joint is to be sweated this strip can be the exact width of the joint. If it is to be done with the soldering iron, make it a little wider. (2) Apply flux to the clean strip. For galvanized iron use hydrochloric acid. For the other metals use the flux given, or a straight zinc chloride solution. Or use commercial soldering paste. Load the soldering iron with plenty of solder and run it along the fluxed area. If some areas do not "take," clean them again with the steel wool or emery cloth, apply more flux and repeat.

TO JOIN THE TINNED PIECES WITH THE SOLDERING IRON

Set them in their correct position. Place them so that pressure can be applied to the upper piece with a stick or the tang of a file. Then pressing firmly down, hold the soldering iron on the tinned part of the metal and melt the joint together. If the joint tends to spring apart, withdraw the iron and keeping up the pressure, allow each section to harden before going on to the next.

Shaping by Hammering in a Hollow Block

Let us for a simple example hammer a small bowl out of copper. Draw a 6″ circle on a piece of 16-ounce copper and cut it out cleanly. Touch up rough places on the edge with a file. Now draw on the copper, concentric circles about ½″ apart. Your hardwood stump should have several saucer shaped depressions carved in the end grain, varying in diameter from ¾″ to 6″ or more.

Method (1): Put the center of the disc over the smallest depression and with your forming hammer drive the metal down into it (Fig. 27A). Move it to the next size and continue to drive the metal down until it takes the shape of the depression. These will guide its placing over the depressions. Move from the smaller to the larger until the copper has been worked right out to the edge. If the metal gets hard

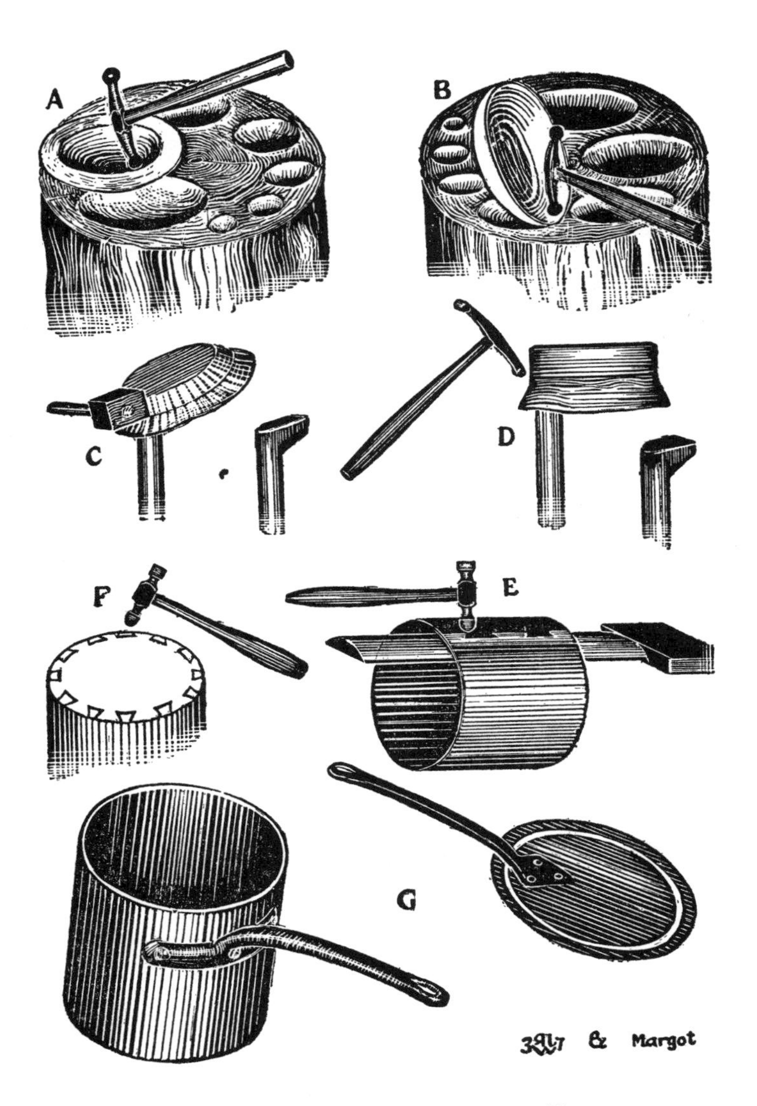

FIG. 27. Shaping Cooking Vessels: Tools.

during this process anneal it. If the bowl is not deep enough at this point gather the edges up higher by the next method to be described.

Method (2): Hold your disc of copper at about 45° to the end of your stump (B) and touching with its edge the outline of the 3″ depression. Drive the ball of your hammer lightly just inside the edge of the disc, forcing it into the depression. Rotate the disc as you go. When you have completed the circle you will have a little turned up rim, surrounding a flat bottom. Start another circle, striking on the next line ½″ further in, and continue all around. By this time the outer edge may be crinkling a little. You must correct these crinkles by striking them with the hammer, filling in the space between the first and second circles with more blows. Your blows should be light but plentiful, rather than strong and few, and this applies to any kind of metal shaping. Continue inwards, using other depressions if advisable, until the whole disc has been worked. Then start again at the outside and draw it up further, repeating until it is as high as you wish, and annealing as often as required. These instructions apply equally to other metals (except, of course, that the annealing does not apply to pewter and aluminum). Some workers prefer to use pointed or rounded wooden mallets instead of the forming hammers, reserving the steel hammers for a final planishing (light hammering to refine the surface and give texture) when the shape is nearly finished.

How to Shape Cooking Vessels

It is time for someone to extoll the virtues of the tinned copper cooking vessels, for they are superior in many ways to the enamelware now in general use. A copper pot is practically indestructible. If hurled forcibly down on a concrete floor, it will receive a denting which can be taken out in a few moments with a hammer or mallet. But let an enamelware pot be dropped ever so gently and it is the recipient of an ugly bruise from which radiates a handsome series of line patterns. Between the lines are little chips of glassy material that eventually loosen and fall out, frequently in the food. A set of copper pots is an heirloom which can be passed from one generation to the next, but few enamel pots see more than two or three years' service, so that in the long run copper pots are much more economical. Add to this the rich and handsome appearance of the metal, and the quality of sentiment associated with things formed intimately by human hands.

Copper cooking vessels with flat bottoms and straight sides are either made in two pieces and brazed, or shaped over the stake from a disc. This is held over the stake with the edge of the stake under a circle that is drawn to represent the bottom. The projection to be turned up should not exceed about 3½". Taller vessels are best brazed. Drive the metal down with a flat faced wooden mallet (C). Keep turning the disc until the circle is complete. The work will now resemble a ladies' straw hat with a floppy brim. Take up a steel hammer (D) and direct your blows against the junction of the "crown" and the "brim," forcing the brim to give way until the shape is all crown, keeping the work constantly turning. True up the bottom by hammering over a wide, flat-topped stake with wooden mallet, and flat-faced steel hammer or by laying the pot on a hardwood or iron block and working with the same tools from the inside.

The two-piece pots are made this way: Take a piece of copper of a width equal to the height of the pot plus about ¾", and of a length equal to 3⅐ times the diameter plus about ½". Make a kind of dovetail jointing in the ends and see that they fit neatly. Clean the metal around the joint, and bend your strip to a circular shape, fitting the dovetails into their proper places. Now holding the piece so that it does not spring apart, put it over a stove-pipe stake or a piece of heavy pipe clamped to the bench; hit the dovetails a few light blows with the ball of your hammer (E) to expand them and hold them from moving. Go along the joint tapping gently until it is fairly well closed. Apply a good coating of the brazing mixture on both sides of the joint, put it to the forge and braze.

File and scrape the surplus spelter from the joint. Put it over the stake and tip the end of the cylinder inwards for about ¾", making a clean flange all round. With a sharp chisel cut another set of dovetails out of this flange. Then put the pot over a disc of copper cut for the bottom and trace the position of the dovetails in the bottom, and fit it, tapping the joint as you did for the side piece (F). Finally apply your paste and braze the joint.

When the joints are scraped clean you may planish the pot all over to true it up and enliven the surface. Then you have the question of handles. Single and double handles may be forged out of copper bars bought as scrap from your junk dealer. For the long handles (G) either make a wooden pattern and have them cast at the nearest foundry or buy some ready-made. Use at least two heavy rivets. They are started from the *inside* of the pot and riveted down over the holes in the handle, on top of a stake.

How to Line Copper Cooking Vessels with Tin

Copper cooking vessels must always be coated on the inside with pure tin to protect them against the acids of the foods, and to protect the foods against the poisonous compounds formed by the acids with the copper. These coatings are applied in this manner: First scour the pot free from all dirt, and boil in it a solution of lye or sodium carbonate to remove all grease. Then after rinsing, pour hydrochloric acid in it and move the acid around keeping it in contact with all the copper until it is bright, then pour out the acid. Now heat up the vessel on the forge and when it is quite hot pour into it a small quantity of pure tin, that you have melted and skimmed in a little ladle. Rub it briskly into the copper with a handful of clean waste. The copper will have to be evenly heated all over to get the best results. If part of the copper refuses to take the coating of tin, rub it with a piece of sal ammoniac (ammonium chloride) continuing to rub with the waste. Add more tin, if necessary, until the coating is smooth, thick and even. The coating should last several years in constant use, but as soon as it does show sign of giving way and exposing the copper, it should be promptly renewed.

CHAPTER 2

On Forging Iron; Making Tools; Grinding and Sharpening Tools

THE smith bending over his anvil in the glow of his bright forge, is always a fascinating picture. Perhaps everyone has envied his power in dominating the strongest of our materials and moulding it to his will. Perhaps it comes as a shock to learn that after all *we* could do such things here and now, if we so willed. Why not? There is nothing impossible in this idea. We could make hardware and fireplace equipment, tables and candlesticks. We could make our own stone-carving tools or wood-chisels. We could attend to small adjustments and repairs we would have done before "if only we had a forge."

The craftsman wishing to work in iron has a good start if he can use the tools and forge of the local blacksmith until he knows just what he needs for himself. Where this is not possible, he must build a forge, or buy one. A solid brick forge built up from the floor, with an arch left in the middle for dumping the ashes from the blower nozzle, is ideal. It should be built against the chimney, and a hole provided in the chimney six inches to a foot higher than the fire. Where the hole is thus set down close to the fire, the smoke is carried away very well without a hood. If a hood is used, it should communicate with the chimney by the most direct route.

On two opposite sides of the forge leave a brick loose in the wall surrounding the fire space, so that if it is necessary to heat up or weld long bars these bricks can be removed to permit the bars to bed down in the centre of the fire. Make the hearth about 18″ or 22″ inside and about 5″ deep. The blower should be purchased and the air ducts set up before the masonry is complete. If so permanent an installation is not desired, get from Sears, Roebuck or Montgomery Ward a portable forge of a size and price that fits your needs. These are provided with blower, hood and legs and are all ready to use.

An anvil that works fairly well for simple, crude work is a piece of railway track mounted on a heavy stump base. Pieces of iron pipe clamped in the vise will do, to bend the curves over. Sooner or later you will want a real steel anvil, and you will wish to have a real blacksmith's vise with a leg that goes down to the floor. The mail order stores mentioned can supply these as well as bar iron of different shapes and sizes, and the tools. It is possible to do good work with few tools, and you should start with a limited selection, adding others as you find it impossible to get along without them. You can make many of the tools yourself. Making special tongs is a valuable exercise in accurate forging and fitting.

You will need a sturdy grinder, an assortment of files, a steel try-square, a steel rule, a pair of steel dividers, a hacksaw, pliers, cold chisels, etc., a drill of some kind, preferably a blacksmith's post drill and bits to fit it, and a collection of hammers. Start with a ball-pein hammer as heavy as you can comfortably swing in one hand, and work up and down in weights from that. You will also need some swages and hot and cold cutters to fit the square holes in the anvil. Use soft coal that does not contain more than ½% sulphur. To start the fire, clear out the cinders down to the air duct outlet, open it from the bottom and dump any dust and cinder into a pail set below it. Then close it ready for blowing. Make out of pine boards a box, shaped like the frustrum of a square pyramid, about six or seven inches high, three inches wide at the top and five inches wide at the base. Stand this with the small end downward over the air outlet and pack your soft coal, moistened with water, around it. Tamp the damp mixture tightly and build it up to the top of the wood, then withdraw the wooden form. Drop into the hole a good handful of lighted shavings and blowing gently, pile on bits of wood, taking care not to build it up higher than the level of the coal. As the fire takes hold of the wood, gather a little fresh coal over the top of it and bank it down, increasing the draft at the same time. When the damp coal begins to coke, you can pile more on top and your fire will be ready.

When you are forming bars of iron for almost any purpose: fire-dogs, tongs, bar-hinges, toasting forks, or what you will, regard the bar as so much raw material. Do not be tempted to leave the bar its full thickness all along its length with a few hammer-marks distributed over the surface to show that it is hand-forged. You are to create things out of *iron* not out of *bars* and you are to make the *iron itself* flow and bend before you in the fury of your driving energy. Be not afraid to reduce the bar to half or a quarter of its weight as you taper it away from the heavier parts.

Watch a blacksmith at work and notice the rhythmic manner in which he relaxes the muscles of his arm, by allowing the hammer to bounce on the anvil for two or three seconds, every few strokes. This is the first thing you should learn. Until you do learn it you will find hammering a very tiring business.

The interest in three-dimensional work of all

kinds is in the alteration and play of all three dimensions. For example, the spiral volute of a snail shell interests us because the diameter of the tube, the radius of the circle and the longitudinal progression all grow larger as we follow them with the eye. But a cork screw or a shock absorber spring move only in their lengths, not in the diameter of the wire, nor in their radii. Hence these seem to us almost as one-dimensional forms, having length, but no other dimensions. We only get the sense of a dimension when we feel that it is being "exercised" by variation.

Notice that old iron workers very well understood the value of this variety in all dimensions. It was perfectly scientific too, because the further away one gets from a fulcrum the less strength is needed in a lever: old hinges, door latches and tongs, for example.

Welding

Obtain some welding compound. This is usually filings of very pure iron that flows easily, mixed with a flux (such as borax). Thicken the parts to be welded by "upsetting," or hammering back against the end. Clean the surfaces and put them in the proper relation to one another, i.e., as you want them to be welded. Sprinkle the welding compound between the pieces and around the joint. Now heat them up to a dazzling white heat, bring to the anvil and strike smartly with the hammer until they are joined. You must hit very hard.

Brazing Malleable Iron

Have the parts neatly fitted and filed clean. Tie them together with iron wire and paint a paste of borax or boric acid, brass filings, and water, into and around the joint. Heat it up until you see the brass run into the joint, then add more brass filings or wire, if necessary. Allow to cool to blackness in the forge before disturbing.

Making Tools

Engraving, Chasing, and Repousse Tools

Use cast steel or drill rod of a suitable size. Anneal it by heating to redness and allowing it to cool slowly. Shape the tools with the file and the grinder or grindstone. Matting and other texture tools for chasing can be made by filing and polishing the square end of a tool, then punching little pits into it with a sharp pointed tool. A contrasting texture tool can be made by hardening the tool just described, and hammering *it* sharply upon the polished end of another annealed blank. Still other textures are made by filing criss-cross grooves on the end of the blank with fine knife files in square or diamond patterns. Harden and temper the tools when they are finished.

Wood-Carving and Stone-Carving Chisels

For anyone who has not worked with tool steel I give this caution: Take care not to overheat your metal by making it really white hot. If it is shooting off glowing sparks, you are burning the carbon out of it and turning it into mild steel. It should be forged at a good cherry-red, and not hammered when it is too cool. When the tools are roughly forged they should be ground, and the teeth filed in the claw chisels with a knife-edge file. Then they are hardened and tempered. Where possible you should obtain and follow the maker's instructions for tempering the particular brand of steel you are using, for some steels are best treated in oil, others in water, others require special solutions. In general you will find this a practical guide.

1. Heat to a bright cherry-red for two or three inches from the tip.

2. Quench about one to one and a half inches in the oil or water.

3. Immediately rub the scale off the quenched part with a piece of abrasive stone or brick, or a bit of emery cloth.

4. Watch the shiny surface of the metal change color as the heat travels down toward the edge from the shank. First it turns a pale straw color, then a brown, then a purple and finally a blue. When the pale straw color has just reached the edge quench the end of the tool again and gently cool off the whole tool by dipping it little by little in and out of the bath.

If the edge is too hard in use and chips easily, heat up the shank of the tool and let a little stronger color creep toward the edge before quenching again, and, similarly, if the edge is too soft, repeat the whole process and quench a little sooner.

After tempering, the tools should again be ground. The Wood-working tools should be ground and polished with emery or pumice stone and finally buffed. Wood handles should be turned on the lathe and brass ferrules cut out of pieces of tubing.

Grinding and Sharpening Tools

High-speed (3450 R. P. M.) direct-drive motor grinders are now available at low prices. These are very useful for rough grinding of tools, and trimming of forged or cast metal. They are difficult to use for fine edge tools because they cut so fast and heat up so quickly. They should be well guarded, leaving only a small segment of the wheel exposed for working. If used much, provision should be made for exhausting the dust with a fan and pipe to a chimney or window. Goggles should be worn. Have a can of water right by the wheel, and dip the tool into it every few seconds to keep it cool. Watch the drops of water on the edge, and when they bubble up you will know that the metal is too hot and needs cooling. If the edge turns black or blue, it means that the metal at that point has been over heated and the temper drawn, leaving it soft. Use the periphery of the wheel for rough grinding and the face or side of it for the finer finishing. The nearer you can get to the center of the face the slower and finer the cutting will be. If you have a variable speed grinder, notice that it cuts much slower at a low speed and there is less heating and less danger of burning the tool. Never try to grind pen knives on a high-speed wheel. The old-fashioned water-fed sandstone wheel is by far the best for fine tools. Its hand-crank or foot pedals can easily be replaced by a pulley and belt drive. See drawing for illustration of the *grinding* angle and *whetting* angle of edge tools. Note that in the grinding solid metal is taken away from the "shoulder" of the tool to make a small angle. The whetting or honing on the oil stone is done only near the edge and makes a larger angle.

FIG. 28. (*a*) *Grinding angle and* (*b*) *whetting angle for planes and firmer chisels.* (*c*) *Rounded shoulders of wood-carving chisels.*

OILSTONES

The double-faced variety of carborundum stone having "coarse" texture on one side and "fine" on the other is useful where there is no grinding wheel, or where—as in schools—the pupils are not to be trusted to use it; otherwise the coarse is not so much used, since the grindstone would do as good a job more quickly. The "fine" side is the same size grit as single face ordinary carborundum stones. These are useful for preparing a new edge after the tool comes from the grindstone or has been worn back by repeated honing on the Arkansas or Washita stones,—fine translucent white stones, the Arkansas being more uniform.

Remove the "feather-edges" of plane blades and flat chisels by pushing them forward on the stone while holding the back of the blade absolutely flat on the stone. To remove the feather edge on gouges and wood-carving tools you will require Arkansas "slip stones" or stones having wedge, round or elliptical cross sections that fit inside the edges. It is useful to have shaped stones for the outside as well, but not necessary, because the tool can be rotated on a flat stone while being moved back and forth. Use pieces of smooth, firm leather for stropping the tools after they have been honed on the Arkansas.

A useful device for honing the inside of the round chisels and gouges is made this way. Turn up a piece of wood on the lathe to a long tapering cone shape, representing from one extremity to the other the range in inside diameters of your gouges. Then rub some liquid glue on to its surface and dust it with emery flour, rolling the cone back and forth through the flour, which should be spread out on a clean paper. When this is dry put it between the centers of the lathe and spin it. Your chisels will quickly respond to a little pressure on this.

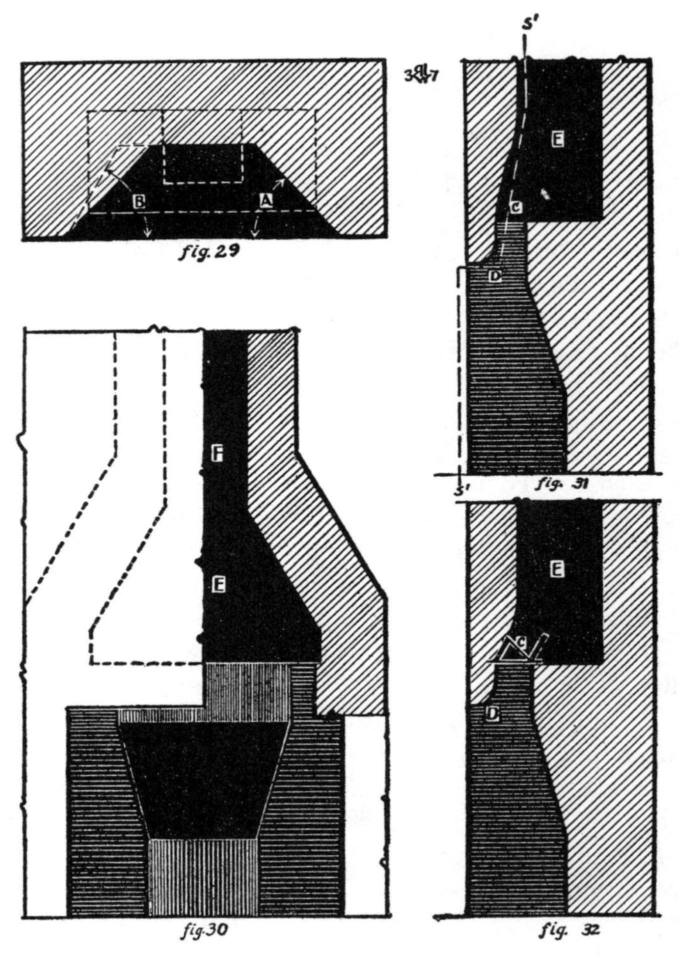

A FIREPLACE HAVING THE CORRECT SHAPE FOR MAXIMUM EFFICIENCY. FIG. 29. *Plan: Large dotted rectangle indicates position of smoke chamber; the smaller one, the flue.* FIG. 30. *Left half: front elevation. Right half: the front wall of the chimney is removed, giving a section as at S′-S′.* FIG. 31. *Correct section (on centre line) for fireplace without damper.* FIG. 32. *Correct section with damper unit.*

Section IV

ON WORKING WITH STONE

CHAPTER 1

How to Build an Efficient Fireplace

Because it is a long time since fireplaces were generally used as the sole sources of heat, the principles governing their correct construction and operation are not generally understood. I propose to show certain simple principles which, if followed attentively, will get you a fireplace that is not only proof against smoking, but also the most efficient in giving forth its heat to the room.

Two Americans—Benjamin Franklin and Count Rumford—contributed more than any other individuals to the early science of domestic heating. Franklin is remembered for his work with the stove which bears his name, and Rumford for his innovations in fireplace design. The fireplaces with which Rumford dealt differed in one important particular from modern ones: they had no dampers. In most houses nowadays there may be long periods when the fireplace is not used at all. During these periods the flue of a fireplace without a damper would be constantly full of warm air rushing from the room to the cold outdoors. It was in order to prevent this waste of heat that dampers were invented.

The fireplace shown has the correct proportions between its parts, and may be relied upon to give efficient heating and no smoke.

CHIMNEY

The chimney should be higher by 2 or 3 feet than adjacent trees, roof ridges, dormer windows, or buildings. The walls of the chimney should be at least 9″ thick and plastered smoothly with mortar on the inside. If less thick than this the brickwork should be protected with refractory flue lining. These come in standard brick sizes.

The net cross sectional area of the flue should not be less than one tenth the area of the fireplace opening. Thus a fireplace opening 36″ wide and 30″ high would require 30 x 36 divided by 10, which equals 108 square inches of flue section. A flue lining 8″ x 18″ (net flue area 109.7 sq. in.) would do; or, better, a 13″ x 13″ lining (net flue area 126.6 sq. in.) A round flue is the most efficient of any for its area, and a square flue is more efficient than an oblong one. These facts should be considered when applying the rule given above.

Beams, joists or rafters should always be framed around a chimney: never built into it nor supported by it. Construction of the latter sort is the commonest cause of fires. If you are altering a house and find potential danger spots such as these, be sure to remedy them before the opportunity to do so is past. Wood should not touch a chimney where the walls of it are less than 8″ thick, but should be insulated from the brickwork with asbestos.

Necessary changes of direction in flues should be gradual, and the flues should not slope to the side more than 30° from the vertical. They should continue the same diameter all the way up from the top of the smoke chamber.

SMOKE CHAMBER (E)

Above the damper a smoke chamber must be constructed, as wide at the bottom as the fireplace itself, and sloping in at about 30° from the vertical to a point exactly above the center line of the fireplace, before the flue proper is started. Under no circumstances start the flue to one side of the center line.

SMOKE-SHELF

The floor of the smoke chamber is formed by a smoke-shelf at the back and the damper in front. The smoke-shelf is very important, for it serves the purpose of receiving occasional down-drafts of cold air and deflecting them forward and upward with the smoke and hot gases.

DAMPER

The damper is always placed as near the front as possible. The plate opens with its forward edge, and is the full width of the fireplace. Use a cast iron damper unit made by a reputable firm, supplied with flanges that rest on the brickwork to make a secure, tight joint. In fireplaces without dampers it is unnecessary to have the opening wider than 4″ at this point.

FIREPLACE

A fireplace is more efficient if it is wider than it is high. It should not be too deep. A deep fireplace wastes its heat on the side-walls and chimney. The nearer the fire is to being built right out in the room, the more heat will be radiated directly into the room.

The side walls of the fireplace are at their greatest efficiency when built at an angle of 45° to the plane of the front, (A. fig. 29). Where smoking is feared this angle may be increased, but it should never be wider than 60° (B), notwithstanding the fact that many fireplaces are built with an angle of this much of more, the result of either ignorance or indifference.

The back wall should curve forward from a point about 14″ above the floor of the hearth, until it meets the bottom of the damper; or, in the case of a fireplace without a damper, until it is just 4″ from the breast of the front wall of the chimney. The throat of the chimney—as this narrowest portion is called—

(C), should be 6″ or 8″ above the bottom of the arch (D) (I speak of it as an arch even if it should have no curve in it).

In order to keep the fire well to the front the arch should not be more than 5″ thick, and it should be rounded off smoothly on its inner edge (D) to provide a smooth ascent for the smoke and to prevent eddying.

The proportion that gives the most efficient radiation of heat, is where the width of the back is equal to the depth, and the depth equal to ⅓ the width of the opening in front, the sides being exactly 45° to the plane of the front. Many people will wish to alter these proportions to make it easier to burn longer logs, but in making concessions to this convenience, they should not forget that if they lower the efficiency of the fireplace by making it squarer and deeper in section, they will have to burn more wood to get an equal amount of heat.

CHAPTER 2

On Stone-Carving; Kinds of Stone; Direct Carving; Sundials

The countryman who is always "fixing his place up" frequently encounters a job where a slight knowledge of stone-cutting would be a great boon.

Stone lintels, keystones, and sills often have to be fitted or altered, a name or inscription carved over a gateway, gateposts decorated or even a headstone lettered for a grave. The drawings (Page 172) show typical stoneworker's tools which any one handy at the forge can make from cast steel stock (Page 157.)

How to Make Straight the Edge of a Block of Stone

1. Draw the line to which you intend to trim your block.

2. Clamp a straight-edge (preferably a flat iron bar) up to your line.

3. With your chipper against the bar, strike it forcibly with your heaviest hammer, and so trim along the whole edge.

How to Trim a Slab

See that the slab is supported under its whole surface. Draw your lines on both sides and very gently tap along the lines with a wide claw chisel, working from both sides until the slab breaks. Trim it with your claw and finish by rubbing down with a large block of carborundum.

Kinds of Stone

Of all the stones used by the colonists for grave markers none has better resisted the weather than slate. The neat inscriptions on slabs of polished slate show even the light scratches that served as guides to the letterers. Artists should consider this when selecting stones on which to work. Slate is a pleasant material to carve. I have carved a full-size portrait relief in gray slate, direct from life, and I found it possible to use a wooden mallet and wood-carving chisels on the stone. To be sure it took the finest edge off the chisels, but not much more than this. A fine toothed claw chisel is good for roughing out. One soon becomes accustomed to the grain. Files, rifflers, sandpaper, and pumice all help in finishing.

The native argyllites used for walks and flagging, and also used for headstones, are second only to slate in resisting the weather. This stone is too uncertain in strength and texture to be used for careful studies and portraits, but it has possibilities for free, fantastic or humorous garden decorations. Slabs set into walls or terraces can be amusingly decorated with simple outline drawings deeply incised and only slightly modelled. It is easy to carve.

The reddish sandstone is next in permanence. Where it was sheltered it stood quite well but did not resist the severity of exposed locations. This stone is easy to carve because it is soft, but wears the tools down fairly rapidly. The color is interesting.

The white native marble, so beautiful to work with, and capable of such subtlety of modelling, is the worst of our stones to weather. Hence it is not to be regarded as the ideal medium for outdoor sculpture. The fact that it is harmless to the lungs of the carver makes it a very desirable stone to use, and methods of preserving it from the weather should be investigated.

The Indiana limestone, Bath and Portland stones, are all easy to carve, but also subject to the weather. They are useful for the beginner to practice on. He can find odd pieces of them lying around the yards of stone contractors, and purchase them for trifling sums.

Direct Carving

In Relief

If the slab is thin support it with plaster against a large piece of stone. Draw the design with pencil or charcoal, and cut the outlines down to the background, clearing it away, and leaving the mass of the figure untouched. Then model it in simple planes, finishing the highest parts first and pushing the finished surfaces toward the background. Note that the dimension represented by the depth of the relief is foreshortened, and the depth of parts such as noses and ears, presented in profile, must be reduced in proportion. Note also that since the relief is a kind of drawing it partakes of the abstract quality of drawing. Emphasize its outlines here and there, by exaggeration of depth to produce a suitable shadow. The light and position in which the sculpture is to be seen in its final location have the controlling influence on such things.

In the Round

Make a block of soap or Ivorite the proportion of the block of stone you intend to carve. Or obtain a small block of talc stone or pyrophyllite of these proportions. The scale should be about 1″ to the foot. Carve a sketch model out of your small block, fitting the figure well into it in a compact pose. For carving use penknives, linoleum cutting tools, or gouges. For the soap or Ivorite use modelling tools as well. Carve the parts of the figure in roughly geometrical abstractions, using cylindrical shapes for the neck, and limbs, ovoid shapes for heads, and for the torso two relatively rigid parts—thorax and hips—joined by the flexible lumbar region.

Do not rely on detail or facial expression for interest. The figure should be sufficiently telling before they are added. The purpose of making the small sketch is to give you the chance of disposing the main masses of your material before you become absorbed in detail. If the first sketch is not a good figure, try another, and do not touch the large stone until the proportions are settled.

Then go to work on the large model with the point, using the sketch for establishing the proportion and disposition of the parts. Finish from the front, pushing the modelled surfaces back with the claw chisels until your work is complete; smoothing further as necessary with the flat chisels, grits and polishing stones.

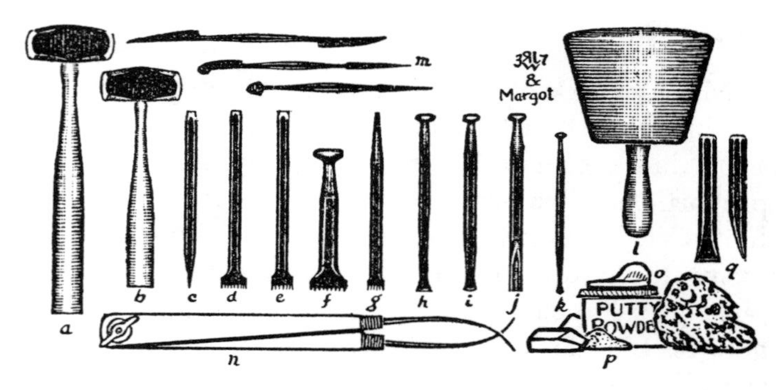

FIG. 33. Stone-Carving Tools: (*a*) *and* (*b*) *1½ and 2 lb. steel hammers;* (*c*) *punch or point;* (*d*) *and* (*e*) *claws;* (*f*) *claw for soft stones;* (*g*) *claw with hardened shank for use with soft iron hammer;* (*h*) *to* (*k*) *finishing tools;* (*l*) *wood finishing mallet;* (*m*) *marble rifflers;* (*n*) *calipers;* (*o*) *felt and wood rubber for polishing with putty powder and water;* (*p*) *pieces of carborundum, pumice and sponge;* (*q*) *chipper.*

Sundials

These are interesting accents for a garden. They may be carved out of solid stone, or the dial with inscriptions may be carved out of a slab and supported upon a pedestal of bricks, rough stone or terra-cotta. I made one sundial for a farm in Great Barrington, Mass. with a direct modelled terra-cotta base and a bronze dial. The Encyclopaedia Britannica gives the mathematical formulæ for calculating the correct angles of horizontal and vertical sundials for any latitude, and these are quite simple to apply if you know a little trigonometry.

CHAPTER 3

Lime Mortars; Cement Mortars; Stuccos and Plasters

The countryman often needs to touch up stonework, brickwork, or plaster walls. He may also build a kiln, forge or furnace or lay a stone terrace. For any of this work he should know how to mix lime, cement and sand. While in the past, bricks in chimneys were laid with clay inside the house and lime mortar outside, today most masonry is laid with cement mortar. If you wish to repair old brickwork, use lime mortar. You can buy slaked or quicklime in 50 lb. paper bags. Cement comes in 94 lb. bags with a volume of 1 cubic foot—a useful hint for calculating concrete batches.

Lime Mortars

To prepare lime mortar, make a box out of wood. Put into it about twice as much water as the lime, shovel the quick lime into it and stir well during the early stage of the slaking.

Instead of using all coarse sharp sand, you may use up to 1/3 of fine soft sand. The small grains of soft sand fill in the voids between the grains of coarse sand. The soft sand is shovelled into the creamy mixture of lime and water and mixed in with the hoe.

This should be still slushy in texture. Cover from the air with soft sand and leave it at least 24 hours. As you need it take the paste out with a shovel and mix with it the right amount of sharp sand (about 1 part lime to 2½ parts total sand).

Cement Mortars

Slake the lime a day or more ahead. Mix dry 1 part of cement and 2½ parts of clean, sharp sand, and form the mass into a crater with plenty of room in the middle. Pour water into it, dragging the sides down as you mix until it is plastic. Then work in ½ to 1 part lime paste. This mortar must be used at once because the initial setting of the cement takes place in 30 or 35 minutes and if this is broken the cement will never develop its full strength.

Stuccos and Plasters

Stuccos and Plasters are made according to either of the methods for mortars. Decorative stuccos for exteriors are cement mortars, often with pure white sand, or colored sands and marble dusts added to give color and texture. Interior plasters usually have little cement except for special purposes. Ready-mixed fibre-plasters are popular with builders because they save time in preparation, but are more costly to use.

Interior plasters are applied in two or three coats. The first coats are made of 1 part lime to 2½ parts sand with sufficient hair to bind it. The hair, obtained from the building supplies dealer as it comes from the oxhides in the lime pits, is matted into lumps filled with dry, dusty lime and dirt. These lumps should be set out on a board in a windy place and well beaten with a stick, keeping to windward so as not to breathe the dust. When the hair is light and fluffy it may be worked into the plaster. A small quantity goes a long way.

The finishing coat for white plaster is made of well-slaked white lime mixed up as required with about equal parts of gauging plaster (plaster of paris), and sometimes a small quantity of clean sand. This must be used soon after mixing. The white "putty" is spread on the wall and wet with water from a brush while burnishing it with the steel float.

FIG. 34.—Stone Carving.

FIG. 35. Block Printing of Fabrics. *Old colonial blocks: (a) and (e) wood blocks; (c) imprint of wood block; (b) and (d) imprints of metal outline strips; (f) metal strips, soldered to a sheet attached to block; (g) charging block in best type of charging tray; (h) printing.*

Section V

ON WORKING WITH COLOR

CHAPTER I

A Brief Essay on Color; Earth Colors; Spectrum Colors

The most satisfying color schemes in the arts are those which tend to approximate the natural ones to which our eyes have been accustomed for untold æons.

Our eyes have been sensitive to color for a long time. It is thousands of generations since human eyes were crystallized in their present form. All of this time we have been gazing at patterns of natural colors; at landscapes built up of soft ochres and umbers and siennas; at brown, dried leaves and the pale bleached gold of winter grass; at the rough, irregular texture of gray bark and pale-green lichens on blunt granite boulders. Our eyes have been accommodated to the mixed vibrations of "impure" colors in the setting of which the "pure" spectrum colors of the flowers sparkle forth like the radiance of jewels.

These pure spots of color are very precious to us and bring us the most exquisite delight. But we are mistaken if we assume that because one square inch of spectrum color brings us joy, a thousand square inches will bring us heaven itself. For we are earth-born folk and could no more withstand heaven's radiance than poor Semele could the glory of Jupiter. The rainbow, which gives us the spectrum colors at their finest and purest is seen as a narrow band against a gray sky.

I suggest the following precepts:

(a) In all kinds of color work, fresco, water-color, tempera, oil, tapestry, batik, block-printed fabrics, or embroidery to have the total area of the earth colors exceed that of the spectrum colors.

(b) To rely to a greater extent upon the emotive qualities of the colors themselves; to exploit, for instance, the stimulating effect of red or the soothing influence of green. Also to render the things of earth with the colors of the earth, and the things of heaven with the colors of heaven (i.e., spectrum colors). I demonstrated this proposition in a mural called "The Judgment of Paris." It consisted of three panels—a long centre panel with Paris and the three goddesses. Oenone, the river nymph, the inconsolable abandoned one, was presented in the left panel modeled first in a monochrome of a powerful earthy red, then glazed with terre verte. This gave the figure a curious sense of coldness and separateness from the glowing landscape background of gold ochre, sienna, etc. with touches of cadmium orange and cerulean blue. In contrast, Helen of Troy was painted in the right panel in warm colors against a cold, unsympathetic background. Paris was modeled in a solid monochrome of sienna with a little raw umber. But each of the goddesses I painted in a pure spectrum color associated with her peculiar qualities. The proportion of the figure was intended to suggest, without other label, her character. You may remember that each of them attempted to bribe Paris. Minerva, whose specialty was wisdom, attempted to show him that wisdom was the most beautiful, and promised to make him wise above all others, if only he would award the golden apple to her as the fairest. Juno, the patroness of commerce and power, advised him that to the mighty belong the spoils, that he who commanded wealth could command wisdom and beauty too. Beauty unquestionably belonged to Venus, the goddess of love, and this unhappy male, faced with the thankless task of bestowing a distinction on one out of three females, would undoubtedly, as an honest man, have awarded the apple to her even if she had not whispered to him that he would one day wed the fairest woman in Greece. Thus, this early beauty contest was not, as is so often represented, a collection of three beautiful women competing on a basis of physical equality. It was, rather, an endeavor to compare the values of three distinct qualities; and each competed frankly on her own merits and abilities. I shall not tell you further what colors I used, lest you think me dogmatic about their associative values; neither shall I reveal the qualities of proportion which distinguished these three. Nevertheless, I believe that few, having seen the three nude figures, would mistake the identities of the goddesses they represented.

For those to whom my definitions of colors may be obscure, and for the assistance of artists unfamiliar with the chemistry of color I here append a list of colors suitable for oil, water-color or tempera painting.

Earth Colors *(Made from natural earths)*

Yellow Ochre	Raw Umber	Indian Red
Gold Ochre	Burnt Umber	Terre Verte
Raw Sienna	Burnt Sienna	Light or Venetian Red

These Also Go With the Earth Colors

Lamp Black or Ivory Black — Oxide of Chromium (opaque)

The above colors are absolutely permanent in oil, tempera or water color and may be mixed together without harmful reaction.

Spectrum Colors

Lemon Yellow (permanent only if made of barium chromate)
Aurora Yellow (the brightest permanent yellow made)
Cadmium Yellow (middle)
Chinese Vermilion or Cadmium Red

Madder Lake	Ultramarine	Cerulean Blue
Manganese Violet	Cobalt Blue	Viridian
Cobalt Violet		

These colors are also highly permanent, subject to the following reservations: Madder fades very slowly if used pure but rapidly if mixed with white or other colors or if applied as a thin glaze. Chinese vermilion darkens slowly in light. The transparent colors—madder, cobalt blue, ultramarine blue, viridoan, and aureoline tend to darken in oil if applied thickly or over a dark ground. This can be avoided by applying them more transparently over a white ground.

CHAPTER 2

How to Paint in Fresco; Preparing the Wall; Preparing the Design; Preparing the Colors; Starting the Work

The increased interest in the beautiful and ancient art of fresco painting is a particularly encouraging sign. The medium is of the highest artistic value, requiring a greater degree of judgment and organizing ability than any other forms of painting. It excels in the fact that it truly becomes a part of the wall, rather than a skin or covering. Architecturally, this is of great importance, only mosaic decoration exceeding it in this respect. It is a medium that is peculiarly adapted to the country, its primitive materials, and simple, direct technique tying it by association with the earth and nature. The clear air of the country is perfect for the preservation of frescoes, their greatest enemy being the impure, acid-laden atmosphere of cities.

Most of the rooms of the country house are potential settings for fresco decoration: in the living room, an over mantel panel; in the kitchen and bathroom, friezes or borders; in the playroom or nursery, the walls may be covered with pictures of animals and legendary characters, painted there by the children themselves.

Roughly outlined, the process is this: on a suitably prepared foundation is laid a thin coating of a mixture of slaked lime and marble dust. The painting, in unchangeable earth colors ground with water only, is applied to this surface. The color, being absorbed by the plaster, becomes imbedded in the calcium carbonate, formed from the slaked lime (calcium hydrate), when it reacts with the carbon dioxide of the air.

Thus the lime reverts to its original condition as limestone (calcium carbonate), from which it was altered when the burning drove away its own carbon dioxide. It takes time for all the lime to be recrystallized, but a skin begins to form at the surface as soon as the plaster is laid, and the colors will not penetrate the film once it is complete. For the artist this is a challenge: he is obliged to finish within a few hours whatever painting he starts.

The fresco painter is not obliged to work on walls. I have had interesting results allowing students to apply the plaster to panels of galvanized metal lath or sheetrock nailed to a rigid frame with galvanized nails. The side of a kiln or chimney is also a tempting place to try out fresco ideas. If they don't work, plaster them over.

For serious work, the wall must be so prepared that it does not get damp; dampness is the worst enemy of fresco (apart from the presence of sulphur in the air, as in smoky cities like Pittsburgh). The wall on which the fresco is to be painted should be separated by an air space from the outside wall.

Preparing the Wall

The ground for the fresco is built up by at least three coats of coarse plaster, the first made of broken brick or pottery mixed with sand and lime; the second the same with a smaller aggregate, and the third coarse sharp sand and lime. The coats are finished rough with the wood float. The total thickness should be greater than an inch. Some of the old Roman and Pompeiian frescoes had grounds up to five inches thick.

Now comes the preparation of the finished surface called the *intonaco*. On its mixing and application depends the success of your work. The best lime is made in the old-fashioned wood-burning kiln, a sort of vertical flue built into a hillside.

Lacking this kind of lime, use the powdered product of commerce coming in 50 lb. paper bags. You can buy either quicklime or slaked (hydrated) lime this way and if you are short of time, use the hydrated. Put it into water and stir it every day for a week, storing it in a tub or crock closely covered from the air. Then put it through a sieve and let it settle. Season for three months or longer and replenish from time to time the layer of water that covers the lime paste, so that the air never has access to it.

Today most lime is burned with coal and there is a danger of sulphur from the coal combining with it in the burning, and later causing efflorescence on the wall. Where you suspect this danger wash the *intonaco* with a solution of barium carbonate as soon as it is laid on the wall, before painting.

When you are ready, mix with the lime paste some pure white marble dust. This is best if the particles are graded in size, the small pieces fitting between the larger ones. Have slightly more than enough lime paste to cement these particles together, allowing for the lime's drying shrinkage. Determine the proportion by experiment. The "marble dust" should be fine, clean grains, like table salt. Some fresco painters use fine white silica sand either to replace part of the marble or all of it. Start with equal parts lime paste and marble. If the mixture sticks to the steel float instead of coming away clean, more marble is needed. Too much lime causes the *intonaco* to crack. Once the mixture is decided on, mix up enough for the whole job and store it in crocks or tubs, covered with water. It cannot be mixed too much. Beat and cut it down with the edge of the trowel many times until it is smooth, unctuous sludge. Before applying it, wet the wall. Spread the *intonaco* with the wood float, and wet it often while you polish it with the steel float. Use small trowels and palette-knives in narrow places.

Preparing the Design

Draw the design to scale: 1½″ to the foot. The figures are to be seen as spots of color, not as individuals, as we are concerned with color relationships at this stage. (With larger scale drawings, you come to the actual work stale, having completed your creation in the preparatory study.) Then make full-size cartoons of all the figures and objects in the study—tone studies in charcoal, sanguine chalk or wash. From these make tracings on thin paper.

Preparing the Colors

The only colors suitable for fresco are earths and neutral mineral colors unaffected by lime. The lakes, owing their color to organic derivatives, are out, of course. Colors that are compounds of sulphur, such as cadmium yellows, vermilion, and ultramarine blue, must never be used.

Colors Suitable for Fresco

The iron reds	*The yellow and brown earths*	*Other colors*
Light Red	Yellow Ochre	Terre Verte (a natural earth)
Indian Red	Gold Ochre	Viridian
Venetian Red	Raw Sienna	Oxide of Chromium, opaque
Red Ochre	Raw Umber	Cobalt Blue
Burnt Sienna	Burnt Umber	Cerulean Blue
		Carbon Black

Purchase the dry colors from a good colorman. Dealers in stucco and plaster specialties sell "Lime proof colors." I used some on a fresco three years ago, and they have stood up very well. One is a bright red, resembling vermilion. Guaranteed by a reputable house for use out of doors, they should be quite satisfactory for fresco.

The colors lighten on the wall after they dry. To judge the tones they should be matched to your color sketch by mixing them *dry* first. For white, dry some of the slaked lime in little cakes, and crush them to a powder. Match the tones to your sketch; grind the powders in water with a muller on a slab; and put them in glass jars labeled to indicate the pigments contained. A stroke of the color should be applied to the label.

Grind the colors for the whole job before you start the painting. Then put out in saucers on a board, enough for the day's work, with a supply of your white remixed to a paste, and a bowl of water. Paint a stroke of each color on the board, to remind you of its effect when dry.

Enlarge on the ground the rough outlines of your composition, from the scale drawing, with squares or diagonals. Paint or scratch them clearly so they cannot be effaced.

Starting the Work

Work from top to bottom, finishing the upper parts first. The plaster and color drip and spoil anything below. Start at either the right or the left upper corner. Lay on to a thickness of about ⅛″ to 3⁄16″, as much *intonaco* as you can cover in a half day; later you will find whether conditions will permit working longer on one area. Apply to the fresh plaster the tracings of your cartoons, and trace the outlines with a blunt point to transfer them to the wall. Then start to paint, with your color sketch and cartoon handy.

Try to paint it at first go. The technique must be transparent like water-color. No overpainting is possible. Do not try to build deep shadows by putting one wash on top of another, for the effect of this is to lighten instead of to darken the color when dry. A satisfactory method is to paint in the highlights first, then surround them with the half-tones and then paint in the shadows. I have experimented with applying the colors with an air brush and compressed air, but I believe that this technique is not adapted for fresco, although it gives beautiful gradations of tone, and clean-cut outlines (by spraying the paint against a cardboard shield, cut to match the outline of the cartoon). The tendency was for the color to pile up on the surface in amounts exceeding the capacity of the *intonaco* to absorb it.

If you make mistakes—and everyone does—cut out the offending part, lay in some fresh *intonaco* and paint it over again. See that you burnish it well into the rest with the steel float, wetting it as much as necessary.

Stop your work where there is a definite change of tone or color in the design, even if this means spreading out into complicated shapes of fresh *intonaco* against uncovered ground. It is impossible to join areas without showing. It is a good plan to indicate part of the background against a figure the better to judge its values. Next morning, trim back the *intonaco* to the outline, and lay in fresh, for the final painting of the background. This keeps the edge of the outline moist so that the new joins better to the old. The trimming should be done slantwise to make a longer joint, with a sharp knife like a surgeon's scalpel.

Work boldly and freely, and forget the slower processes of oil painting. And have a good time doing it.

CHAPTER 3

ON THE ART OF DYEING WITH OLD VEGETABLE DYES; PREPARING AND DYEING WOOL; DYEING LINENS

DEERFIELD, Massachusetts, was famous a few years ago for its Society of Blue and White Needle Work, producing beautiful patterns of applique and embroidery from materials dyed with vegetable colors. For many years these were gobbled up by a demand beyond the supply. Several women were kept working under the direction of Misses Margaret Whiting and Ellen Miller following their designs with colored yarns and fabrics dyed on the kitchen stove. Some of the hangings display an engaging ingenuity in the manner in which a pattern or picture was formed by sewing together pieces of cloth of different colors and textures and by the exploitation of differences of character and scale in the stitches.

Miss Miller died in 1928 and Miss Whiting, owing to failing eyesight has been unable to carry on the work. It seemed a great pity that all of the craft and lore on vegetable dyeing should be lost to others interested in following up the craft.

Having already worked with Miss Whiting's material, I suggested recording some of it in this book. She fell in at once with the idea, giving me access to the notes left by Miss Miller, who did all of the colors except indigo, Miss Whiting's specialty. These were jottings designed for her own use and not easy to decipher. I spent many days going through them, experimenting with formulæ, and examining "swatches" (samples of dyed material with their appended notes). Tested by exposure to south light, and filed with unexposed swatches of the same dyeing they show the relative fastness of the different methods.

Dyeing is an *art* not a science. As in other arts, there are several ways of arriving at the same result. None will produce good results without thought, care and patience. The following formulæ should serve as starting points for your own experiments rather than fixed and final proportions. The quantities of water should be sufficient to cover the material to be dyed, the weight of which forms the basis of calculation for the dyebath.

I give in the following pages a good deal of information picked up elsewhere, as well as a general chemical background and some formulæ of my own.

Mordant

Most dyes must be applied or set in the presence of a mordant. A few mordants are: alum (potassium aluminum sulphate), cream of tartar (potassium tartrate), copperas (ferrous sulphate) also known as green vitriol, potassium bichromate, tin crystals (stannous chloride), tannic acid or barks or wood that contain it.

Dyestuff

The dyestuff is the substance which colors the fibres. The same dyestuff may give different colors with different mordants. For example, linen mordanted with alum and dyed with madder is red; prepared first with lime and acetate of lead, rinsed in oxalic acid, and dyed with madder with potassium bichromate it is a golden yellow; if the potassium bichromate is omitted, the color is a reddish purple.

Washing

Thorough washing and rinsing are indispensable. Clean work is impossible if this is neglected.

Kinds of Fibre

There are two chief kinds of fibre: Animal, including wool and silk; and vegetable, including cotton and linen and the modified vegetable derivatives—rayon, celanese, etc. Each of these requires its own technique, but the animal fibres are easier to dye than the vegetable, and the natural vegetable fibres easier than the artificial.

Preparing and Dyeing Wool

Wool as sheared has a great deal of dirt and grease in it, necessary to remove before spinning or dyeing. Nowadays this grease (known as Lanolin) is recovered, refined, and sold to the chemical trade.

The people washing wool on a small scale seldom attempt to recover the lanolin.

Throughout the Scottish Highlands they clean wool with stale human urine, one part urine to four parts water. The bath is no hotter than the hand can bear—about 101° F. The Wool is squeezed and worked about by hand until it is clean, then thoroughly rinsed. Although nothing leaves the wool softer than urine, many people will prefer to wash it in a solution of ammonia or soft soap and hot water. Do not use washing soda, it makes the wool harsh. The wool loses about 20% of its weight.

LICHENS

The simplest vegetable dyes are the Lichens, used in the Highlands of Scotland and parts of Ireland. The well-known Harris Tweeds are dyed with lichens. The wool is dyed before spinning, different shades being carded together to make mixtures. Its characteristic smell is from the lichens and sea-weeds used in the dyeing, and the peat smoke of the hearths near which it is dried.

The *Parmelia Saxatilis* and *Parmelia Omphalodes* (lichens) are gathered off the rocks in late summer, dried, and boiled up with water. This, after cooling, is heated up again and the wool boiled in it for all brown shades until dark enough. I am advised that these are not native to America; but there are others, such as the *Usnea Barbata*, found in Pennsylvania on old trees. It dyes wool orange.

The *Lecanora tartarea* is a rock lichen, abundant in Scotland and Northern Europe, used to dye a red. One method of preparation: crush the dry lichen to a powder in a mortar with a little kelp, and add stale chamber lye. Stir occasionally for some weeks. Knead the paste into balls with a little lime. To use: Soften to a jelly in warm water and fold it into the cloth layer for layer. Boil the whole in water with a little alum.

LOGWOOD BLUE

A good fast blue, easier to use than Indigo: Put 5 lbs. wool in water containing 2½ oz. bichromate of potash. Boil for 1½ hours. Rinse and dry. Boil the wool for an hour with 1 lb. logwood chips in fresh water. Rinse well and dry.

LOGWOOD GREEN

Boil wool dyed blue as above in the hot heather liquor described below until it is the color you wish.

HEATHER YELLOW

Boil green heather tips not yet flowered—about as much as you have wool to dye, in water for half-an-hour. Boil the wool first for an hour in a solution of alum—(1 oz. to a gallon of water). Golden rod and nettle roots and leaves give similar shades. For Moss Green, mordant the wool with the bichromate solution given under Logwood Blue, and then dye it in nettle liquor. The Indians used the Canadian three-leaved Hellebore (*Helleborus trifolius*) for dying skins and wool yellow. Holly gives yellow with alum and green with ferric chloride. Smartweed also gives yellow with alum.

RED

Boil for an hour ½ the weight of wool being dyed, the roots of Ladies Bedstraw in water to cover. Boiled first in the alum solution it gives an orange-red, with the bichromate solution a crimson. *Beet Red:* Mordant wool with alum and boil in beet juice.

MADDER-RED

A reliable method is as follows; for 6 lbs. woolen fabric dissolve 1¼ lbs. alum, 5 oz. tartar, 5 oz. tin crystals in water. Boil the goods two hours, remove, cool and drain over-night. Stir 4½ lbs. madder into fresh water. Enter fabric at 120° F. and raise temperature to 200° in one hour. Handle, rinse and dry. Highlanders mordant the wool with ¼ pound of alum to a gallon of water. Then bring to the boil very slowly; and simmer for an hour in bath with 3 lbs. madder to 5 lbs. wool.

FAWN

Boil yellow flag roots with the wool for an hour. For plum color, add a little copperas just before the dyeing is finished.

BLACK

Boil for 2 hours 1 lb. wool with 1 lb. of twigs and sappy bark of alder, oak, butternut or walnut, or the green walnut husks, and 4 oz. logwood chips, for a reddish brown. Add ½ oz. copperas for fast black.

BROWNS

Boil wool with barks and twigs mentioned above or with leaves, twigs and bark of sumach, maple, hemlock, hickory, etc. used without a mordant, or with alum or acetate of alumina.

Dyeing Linens

Many formulae given for wool will dye linen also, but the vegetable fibres usually require a more elaborate technique. First a few typical solutions and materials:

LIME WATER

Stir ¼ lb. slaked lime in a gallon of water. Allow to settle. Use clear liquid.

ACETATE OF ALUMINA (Red Liquor)

Dissolve 2½ lbs. alum in 1 gallon boiling water, then add 3 lbs. acetate of lead. Dissolve, let settle, and use the clear liquid.

ASTRINGENTS

Nut-galls (excrescences on the leaves and twigs of oak trees) or sumac leaves, twigs and buds of flowers, boiled in water, make an astringent bath that imparts a yellow color to the fabric.

CATECHU (CUTCH)

Catechu is the extract from the wood of the *Acacia catechu*. Like indigo, its color is produced by oxidation, but unlike indigo its oxidation is very slow. In the East the natives produce their browns by natural oxidation. In the West it is accelerated by chemicals or steam.

LOGWOOD

A heavy tropical wood used in the form of chips, raspings, or liquid extract. Tie the chips in a bag and boil with the material, or boil and strain before the linen is entered.

DYEING DIRECTIONS

CUTCH BROWN

For 1 lb. linen dissolve 1 oz. cutch in boiling water. Work goods for 15 to 30 minutes and pass through a hot bath of bichromate of potash. Rinse and dry. (May also be boiled in soap suds after drying to further fix the color.) Copper sulphate with ammonium chloride in the oxidizing bath will give a different hue to the dye. Lime water can replace the bichromate.

CUTCH GRAY

Boil for one hour 1 lb. linen, previously dyed cutch brown, with ½ oz. nut-galls in water. Air, and dye ½ hour in cold solution of 1 oz. copperas. Air, rinse and dry. The linen is more gray if entered directly in the gall-bath without rinsing the bichromate from the cutch dyeing.

LOGWOOD BLUE ON COTTON. An intense blue.

Work the goods in a bath of logwood liquor with the addition of acetate of copper, raising the bath from room temperature to 122° F. in one hour.

ROSE RED

Mordant 1 lb. linen with acetate of iron and acetate of alumina. Dissolve in water ½ oz. madder extract and ½ oz. bichromate of potash. Keep linen in warm dyebath on back of stove for two or three hours and leave in dye to cool overnight. Boil half hour in soapsuds. Rinse.

A MEDIUM ROSE

Turn 5-10 minutes in hot solution of alum and lye. Wring out and dry. Dye in madder and bichromate of potash. Keep madder dyes below boiling point.

FUSTIC GREEN

Boil 1 lb. linen one hour in water with 3 oz. acetate of lead. Lift and dry. Boil one hour with fresh solution of 1 oz. fustic extract, ½ oz. copper sulphate, small amount logwood decoction. Air goods, then add to bath ½ oz. bichromate of potash ½ teaspoon oxalic acid. Boil ½ hour. Rinse again through hot acetate of lead. Then pass directly through hot suds, and rinse in cold water.

INDIGO VAT DYEING

Triturate 1 lb. indigo to an impalpable powder with warm water. Dissolve 2 lbs. copperas in hot water. Slake 3 lbs. lime in cold water; add the three solutions in above order to a crock filled with cold, soft water, stirring each vigorously as it is poured in. A weak bath and several successive dippings give evener color than a strong bath with few dippings. Cover closely and leave about ¼ hour, when it should appear a bright pea-green. Stir again thoroughly, taking care not to stir air into the bath, for this will make the color uneven. Cover, and leave over night.

Make a net of cheese cloth bound to a wooden hoop with legs weighted with stone (not metal) to prevent floating. This is to prevent the cloth from being in contact with the sludge at the bottom of the crock. It is taken out of the crock before stirring and replaced after the settling.

To Use: In the morning remove the cover and skim the scum from the surface. The scum is called "flowers of indigo" and is to be saved in another crock for replenishing the bath later. Estimate the quantity of indigo in the scum after it has accumulated, and add to it sufficient copperas and lime solutions. The cloth is wet in hot water, allowed to cool without wringing, and entered gently but quickly into the bath. Air bubbles may cause light spots. Work gently in the liquor with the hands (which should be covered with rubber gloves, if much dyeing is to be done) for two to four minutes, keeping it below the surface of the liquid. When removing it from the vat gather it into the hands so as to exclude as much air as possible and lift it instantly into a pail of clear soft water. Work it under the surface and then hang it in the shade, preferably in the wind, to air. After about half an hour the color will have developed to a blackish or muddy blue and you may repeat the dipping, rinsing and airing until the color is sufficiently deep.

RINSING AND SOUR BATH: Rinse till no color comes out. Remove the lime from the fabric by rinsing thoroughly in a "sour bath" made by mixing a few drops of sulphuric acid into a pail of water. Rinse well in clear water afterward. Dry in the shade.

VAT: The vat keeps indefinitely if protected from the air when unused, and cared for by stirring, skimming and occasionally removing the sludge from the bottom. This may be thrown away.

ANOTHER METHOD: Fill dye-vat with water at 120° F. Add sufficient caustic soda to make the bath decidedly alkaline. Triturate indigo with warm water and stir into the bath. Stir sodium hydrosulphite in until the color of the bath changes from blue to green and then to greenish yellow. This bath is used just like the other except that it is better to keep it close to 120° F.

CHAPTER 4

On the Art of Printing on Fabrics With Wood-Blocks; Dyeing Medium; Printing; Steaming; Further Experiments

The art of textile block-printing is so simple, and capable of such variety that it is a pity it is not widely practiced here, as in more primitive countries, as a home craft. You can give your bed-spreads, curtains, linens, and dress goods great individuality and distinction by printing on them your own designs in color.

Examine Colonial hand-blocked textiles for types of design suitable to this technique. See page 176. Try a simple repeating border first. Cut your block (as directed on page 76).

Dyeing Medium

No *pigment* applied to the outside of a woven material becomes a part of it in the sense that a genuine *dye* becomes a part of the fibre of which the fabric is woven. For this reason all schemes for applying printing ink, oil-paint and I-don't-know-what other foreign matter to the surface of fabrics, should be returned with thanks to their originators; for I should as soon expect an honorable man to compound a felony, as one who prizes his esthetic integrity, to compromise with such unsound schemes as these.

Dyes for printing require thickening to prevent their running. Commercially, Irish Moss, Blood Albumen, starch, etc. are used, but for work on a small scale nothing is better than Gum Tragacanth, from your druggist. Add ½ oz. to one quart of water. After 24 hours stir it vigorously with an egg beater. Soak 24 hours more and stir again adding a few drops formaldehyde or carbolic acid.

You could make your first experiments by dissolving in hot water part of a package of one of the dyes made for home dyeing and stirring it a little at a time into your thickening, and then try printing with this. You could write the manufacturer of this dye telling him what you are doing, and ask for further technical suggestions. Afterward, if you want to tackle the problem more professionally, you should write Du Pont, General Dyestuffs, or other manufacturers for sample swatches of dyes specifically made for printing and also for detailed instructions. Formulation of these printing thickenings and the subsequent mordanting or heat treatment to fix the dyes is a formidably complex subject, which could easily scare you off if you were exposed to it suddenly: it would be wise when writing to be frank about the limitations of your knowledge and equipment, and specific about the type of textile you are working with and your manner of applying the color.

Printing

Lay in a shallow tray or cake tin larger than your block a piece of thick, soft felt. Pour sufficient thickened dye into the tray to saturate the felt, working it about until the dye is evenly distributed. Add dye to the pad as the block takes it out for printing.

Press the block gently on the pad a few times until it is evenly coated with color. Press firmly on a sample of cloth on a well padded table. Practise until you can make perfect impressions, before printing the final piece.

To hold the dye more evenly, varnish the printing face of the block and sift cotton flocks (fine lint obtained from textile mills) over it while the varnish is sticky. Repeat until a felty surface results.

Steaming

Thoroughly dry the printed pieces. Lay them on clean newspaper (10 sheets thick) and roll up carefully. They must not overlap or touch each other. Suspend this roll in a netting across the top of a washboiler, well out of reach of splashing from the water below. Cover with old towels or blankets, and steam at least an hour. Finally rinse well in cold water and dry in the shade.

Further Experiments

Now try all-over patterns in one color. Later you can experiment with two-color borders. Print all of one color first, letting it dry before starting the next. An endless variety of color-schemes is possible with two or three blocks. Interesting variations are made by printing with a mordant in clear thickening and then dipping in an unmordanted dye. See page 191. Only the printed part will take the dye and the rest will rinse out. Another scheme is to print with a discharging chemical (sodium hydrosulphite) on a previously dyed fabric. In this case the printed part will bleach out. In printing make use of a T-square, straight-edge or even a carpenters' chalkline for lining up the impressions as you print borders or continuous patterns. Do not be afraid to experiment with new ways of applying blocks.

Bibliography

As everything in this book has been written "right out of my head," I list the following books, not as authorities for my statements or opinions (for which I take full responsibility), but as books which I have liked and can recommend.

The Potter's Craft, C. F. Binns. (D. Van Nostrand Co., N. Y.)

A Method for Creative Design, Adolfo Best-Maugard. (Alfred A. Knopf.)

Stage Scenery and Lighting, Samuel Selden & Hunton D. Sellman. (F. S. Crofts & Co., N. Y.)

The Painter's Methods and Materials, Prof. A. P. Laurie. (Seeley, Service & Co., London)

Wood-Carving Design and Workmanship, George Jack. (D. Appleton, N. Y.)

Modeling and Sculpture, Albert Toft. (Seeley, Service & Co., London.)

A Potter's Book, Bernard Leach. (Faber & Faber Ltd., London)

Autobiography, Benvenuto Cellini.

Treatise on the Goldsmith's Art, Benvenuto Cellini.

Dyeing and Calico Printing, Sir William Crookes.

THE COMPLETE BOOK OF POTTERY MAKING, John B. Kenny. (Chilton Co., Book Division, Philadelphia)

HOME LIFE IN COLONIAL DAYS, Alice Morse Earle. (Macmillan Co., N. Y.)

A WOODCUT MANUAL, J. J. Lankes. (Henry Holt & Co., N. Y.)

EVERYDAY LIFE IN THE MASSACHUSETTS BAY COLONY, George Francis Dow. (The Society for the Preservation of New England Antiquities, Boston.)

THE ELEMENTS OF DYNAMIC SYMMETRY, Jay Hambidge. (Brentano's, N. Y.)

SIMPLE COLONIAL FURNITURE, Franklin H. Gottshall. (Bruce Publishing Co., New York, 1931.)

METAL ART CRAFTS, John G. Miller. (D. Van Nostrand Co., Inc. New York, London, Toronto.)

THE COLONIAL AND FEDERAL HOUSE, Rexford Newcomb. (Lippincott Co., Philadelphia, 1933.)

IRON AND BRASS IMPLEMENTS OF THE ENGLISH HOUSE, J. Seymour Lindsay. (London and Boston, The Medici Society.)

MAKING POTTERY, Walter de Sager. (The Studio Publications, N. Y.)

WOOD ENGRAVING, Bernard Sleigh, R.B.S.A. (Sir Isaac Pitman Sons, London.)

WOOD ENGRAVING OF THE 1930's: Reviewed by Clare Leighton. (The Studio Publications, N. Y.)

COLOUR BLOCK PRINT MAKING FROM LINOLEUM BLOCKS, Hesketh Hubbard. (Breamore near Salisbury, England.)

LINOLEUM BLOCK PRINTING, Ernest Watson. (Milton Bradley, Springfield, Mass.)

HAND-BLOCK PRINTING ON FABRICS, T. J. Corbin, A.R.C.A. (Pitman, London, 1934.)

PUEBLO POTTERY MAKING, Carl E. Guthe. (Yale University Press, New Haven.)

CERAMICS FOR THE ARTIST POTTER, F. H. Norton.

WOODCUTS AND WOOD ENGRAVINGS: HOW I MAKE THEM, Hans Mueller. (Pynson Printers, New York.)

Manufacturers

Here follows a list of manufacturers from whom may be obtained materials mentioned in this book.

Pyrometric Cones:

The Standard Pyrometric Cone Co., 1445 Summit St., Columbus, Ohio.

Woodblocks:

J. Johnson & Co., 125 Fulton St., New York.

Inks and brayers:

American Typefounder's Co., Boston and New York.

General Printing Ink Corp., Norwood, Mass.

Paper for printing:

Japan Paper Co., 109 E. 31st St., New York; 453 Washington St., Boston, Mass.

Thomas Nast Fairbanks, 373 Fourth Ave., New York.

Vitreous enamels for metals:

Thomas E. Thompson, 1417 Central Ave., Wilmette, Ill.

Prepared pottery clay:

Western Stoneware Co., Monmouth, Ill.

Electric kilns and wheels, glazes, cones, clays:

American Art Clay Co., Indianapolis, Indiana.

Tools for stone-carving:

Vermont Marble Co., Rutland, Vermont.

Tools for metal work:

Allcraft Tool & Supply Co., Inc. 15 W. 45th St., New York 35.

Anchor Tool & Supply Co., Inc. 12 John St., New York 38.

Kilns and General ceramic supplies:

Standard Ceramic Supply Co., 1466 River Ave., Pittsburgh 12.

American Art Clay Co., Indianapolis, Ind., Toronto, Canada.

Wenger's, Stoke-on-Trent, England.

B. F. Drakenfeld & Co., Inc., 45 Park Place, New York 7.

Hammil & Gillespie, Inc., 225 Broadway, New York 10007.

Kiln Furniture:

Loutham Mfg. Co., E. Liverpool, Ohio.

Norton Co., Refractories Div., Worcester 6, Mass.

Prepared Frit:

Standard Ceramic Supply Co., Pittsburgh.

Pemco Division, the Glidden Co., Baltimore 24, Md.

Pug mill (extruder):

Fate-Root-Heath Co., Plymouth, Ohio.

Insulating Firebrick:

Johns-Manville, 22 E. 40th St., New York 16.

Thomas C. Thomson, 1539 Deerfield Rd., Highland Park, Ill.

Blen-tex, Bell Operations, 157 Virginia Ave., Chester, W. Va.

INDEX

Index